Medical physics

Second edition

Jean A Pope

BA (Oxon)
Plymouth College

The *Options in Physics* series

Series editor: Jean A. Pope

Medical Physics
Rotational Dynamics
Nuclear Physics

Also by Jean A. Pope:

Comprehension and Experimental Analysis in A-level Physics
Heinemann, 1973

Preface to the second edition

In this introductory text on Medical Physics, the intention throughout is to apply the fundamental principles of physics to describe body functions and the instruments used to monitor them. The material selected for inclusion reflects the requirements in particular of the Joint Matriculation Board's A-level option on Medical Physics, but it is also intended for anyone embarking on a career within the medical profession or allied fields.

In order to provide a more complete coverage of the University of London's A-level option on Medical Physics, new material has been added, particularly in the sections on ultrasound and nuclear medicine.

The book is divided into three sections as follows:

The physics of the human body

The body's ability to move and perform work is analysed using the basic concepts of force and energy. The human response to light and sound energy is then considered in some detail.

Biomedical measurement

This section describes the monitoring of various physiological parameters, such as temperature and pressure, using specialized transducers and systems employing, for example, radiotelemetric, ultrasonic and fibre-optic techniques.

Ionizing radiation

The production, properties and detection of the different radiations are discussed and the elements of dosimetry presented. Both diagnostic and therapeutic applications of the radiations are dealt with: emphasis is on the former, with a detailed treatment, in particular, of radioactive tracer studies.

I should like to thank Roger for his constant support throughout the preparation of this book, our parents for their help and encouragement, my friends for their assistance with typing and the Joint Matriculation Board [JMB] and the University of London School Examinations Board [L] for their co-operation and permission to reprint past questions. My thanks are also due to Christopher, Vicky, Nicholas and David for their continuous demonstration of energy expenditure.

J. P.

Contents

PART II BIOMEDICAL MEASUREMENT

5 Electrical conductance   67

6 Temperature measurement   75

7 Pressure measurement   89

8 Radiotelemetry   100

PART I
The physics of the human body

1 | Expenditure of energy

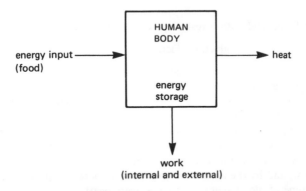

Figure 1.1 The human machine

Energy input

The chemical energy contained in food provides the body's energy input. There are three basic foods or fuels: carbohydrates, fats and proteins. The process of digestion reduces (a) carbohydrates to glucose or closely related monosaccharides; (b) fats to glycerol and long-chain fatty acids; and (c) proteins to amino acids. These organic compounds are then oxidised or 'burnt' using the oxygen inhaled through the respiratory system. As the body temperature is not high enough to burn fuels directly, enzymes are needed to act as catalysts in the different oxidation processes. These all follow the same pattern:

$$\text{fuel} + \text{oxygen} \xrightarrow{\text{catalyst}} \text{water} + \text{carbon dioxide} + \text{energy}$$

However, the exact metabolic pathways and quantities involved vary considerably. The process is generally known as aerobic combustion or respiration.

Table 1.1 shows the approximate quantities of energy liberated by the complete combustion of 1 kg of each of the three fuels. However, since incomplete digestion, absorption and utilisation leads to losses through excretion, the actual energies available are somewhat smaller, as the table shows.

The energy-releasing fuel mixture 'burnt' in the body consists mainly of carbohydrates, with some fats and a little protein. Proteins are rarely used as a fuel except in cases of extreme starvation or excess food intake.

When there is insufficient oxygen available to sustain normal aerobic respiration, as is likely to occur, for example, during heavy exercise, the alternative process of anaerobic respiration occurs. Here, glucose in the muscles is converted

Table 1.1 Body fuels

Fuel	Heat of combustion (MJ/kg)	Available energy (MJ/kg)
Carbohydrate	17.2	16.7
Fat	39.4	37.7
Protein	23.4	16.7

directly into lactic acid with a resulting release of energy:

$$glucose \rightarrow lactic\ acid + energy$$

As lactic acid is a poisonous substance, anaerobic respiration can only continue for a limited period before muscle fatigue occurs and pain is experienced. Furthermore, the process is relatively inefficient, yielding only about a twentieth of the energy liberated by aerobic respiration.

During recovery, the muscle continues to use oxygen at a fast rate, to make up the 'oxygen debt' and oxidise the lactic acid. The oxygen supply can be increased by panting (increased rate and depth of breathing), by local massage to increase the blood flow, and by the automatic responses of increased pulse rate and stroke volume (volume of blood pumped at each heart beat).

Energy storage

Energy is stored in the body in a chemical form. The two major reserves are the stores of fats and carbohydrates.

Fats

Fat is stored in the cells of specialised connective tissue, known as adipose tissue, which is found in all parts of the body but mainly under the skin, around the kidneys and in the abdominal cavity. It is the body's major energy store, comprising on average about 10 per cent of male body weight and 25 per cent of female body weight. When oxidised, fats produce about twice the quantity of energy released from an equal weight of carbohydrates (see Table 1.1).

Carbohydrates

After being processed by digestion, carbohydrates may be stored in the liver and muscle cells in the form of glycogen. When energy is required, glycogen is broken down by enzyme action into units of glucose ready for use in aerobic or anaerobic respiration. The energy thereby released may either be utilised directly or further stored in high-energy molecules, mainly adenosine triphosphate (ATP) in the muscles. ATP can later return to the lower energy form adenosine diphosphate (ADP) with a corresponding release of energy.

In fact, only 1 per cent by weight of glycogen is present in resting muscle, so this store is soon exhausted. Thereafter, glucose supplied by the blood provides the source of energy.

Basal metabolism

Energy is needed to maintain the basic body functions, for example:

(a) to build up the proteins required for making new cells in growth and repair;
(b) to produce the muscular movements necessary for vital internal activities such as respiration, blood circulation and digestion;
(c) to maintain the ionic gradients between cells and body fluids;
(d) to manufacture the secretions of the glands;
(e) to transmit nerve impulses throughout the body.

The utilisation of energy by the body when it is completely at rest is described as basal metabolism, and the rate at which energy is used in this steady state is known as the basal metabolic rate. For an average young man, the basal metabolic rate is about 85 W, whilst an average young woman has a slightly lower rate. Table 1.2 shows the approximate distribution of energy expenditure throughout the resting body.

Table 1.2 Energy expenditure

Organ	Rate of energy consumption (W)	Percentage of basal metabolic rate
Liver and spleen	23	27
Brain	16	19
Skeletal muscle	15	18
Kidney	9	10
Heart	6	7
Remainder	16	19
Total	85	100

Ultimately, all the energy expended performing this internal work is degraded into heat, which is then used to maintain a steady average body temperature of 37°C.

Measurement of energy expenditure

Food intake cannot be used as a measure of energy expenditure over any given time as the body can store energy. Other methods have therefore been devised and these fall into two categories as described below.

Direct measurement

A human calorimeter is used to measure the total heat output from a subject. The first of these calorimeters, named the Atwater chamber after its designer, consisted of a room in which the subject lived for several days during which time all outputs and inputs were monitored for heat exchanges. The heat produced by the subject was determined from the rise in temperature produced in water circulating in pipes through the chamber. Any external work done by the subject and subsequently stored in such a chamber (for example, raising weights) is added to the heat output to give total energy expenditure. External work not stored (for instance, using an exercise bicycle) and internal work need not be added as these ultimately degrade into heat and appear in the heat measurements.

A subject's total energy expenditure may thus be monitored whilst he is engaging in various activities.

Indirect measurement

Since energy is made available for use by the oxidation of various fuels, a measurement of oxygen consumption must provide an estimate of energy expenditure. However, as the different types of fuel release different amounts of energy for the same oxygen consumption, a knowledge of the metabolic mixture is theoretically required before energy expenditure can be estimated. In practice, a knowledge of the respiratory quotient, RQ, given by:

$$RQ = \frac{\text{volume of oxygen consumed}}{\text{volume of carbon dioxide produced}}$$

is sufficient to assign a definite joule equivalent for each cubic metre of oxygen consumed. For example, an RQ of 1 corresponds to about $21 \, \text{MJ m}^{-3}$ of oxygen and indicates that the fuel consists entirely of carbohydrates.

Various methods are available for monitoring inspired and expired air. One of the more sophisticated devices, known as the integrating pneumotachograph, or IMP, supplies air of a constant known composition to the subject via a mask similar to a pilot's oxygen mask. The expired air is then passed to a flowmeter which records the total volume of air flowing through it. A small sampling pump withdraws a sample of air at regular intervals for storage and later analysis.

The information obtained in such experiments is of great use in sport, in military training, in the design of equipment such as rucksacks, tools and machinery, and in the treatment of certain conditions such as obesity.

External or mechanical work

In addition to the energy consumed to keep the body 'ticking over', further energy is necessary for the performance of any 'external' work. This includes even the simplest tasks like sitting and standing, which require extra muscular activity to maintain posture.

Table 1.3 shows the approximate rate of energy expenditure during some common activities.

Energy and efficiency

The efficiency of a machine is defined as the ratio

$$\frac{\text{useful power output}}{\text{power input}} \times 100\%$$

Different organs of the body have different efficiencies, ranging from less than 1 per cent for the kidneys, to 25–30 per cent for the muscles and heart during exercise.

Table 1.3 Energy expenditure during exercise

Activity	Power consumption (W)	Factors affecting power consumption
Resting (basal metabolism)	60–100	sex; weight; amount of fat
Sitting or standing quietly	80–175	sex; age
Walking	150–500	weight; walking speed; walking surface; gradient
Light sports, e.g. cricket, golf	150–350	various
Carrying loads	175–600	load; speed
Moderate sports e.g. gymnastics, swimming, tennis	300–525	various
Climbing stairs	400–850	weight; vertical speed
Heavy sports, e.g. rowing, squash	400–1400	various
Running	700–1400	speed; gradient
Extreme activity, e.g. cycle racing	≈ 1600	

The gross efficiency of the body lies between these limits, being greatest during strenuous exercise when basal metabolism accounts for only a small proportion of total energy expenditure. During moderate exercise, the body can develop power with an efficiency of about 20 per cent, a figure which compares favourably with man-made machines.

Exercise 1

1 What is meant by *basal metabolism?* Describe the energy conversions occurring in the body during the following activities: (a) sleeping, (b) eating, (c) running uphill.

  If the average basal metabolic rate is 7×10^6 J per day, how many people at rest in a room would emit heat at a rate equivalent to a 1 kW fire?

2 Explain why a long-distance runner:
  (a) may eat glucose tablets during the race;
  (b) pants as (s)he runs;
  (c) does not wear heavy sweaters while running to prevent energy loss;
  (d) may suffer from muscle fatigue and require massage;
  (e) feels hungry after the race.

3 Why does the body generate about the same power as an electric light bulb, even when asleep? Where does this energy go?

  It is found that about 1 J of mechanical work is done during four complete (in and out) breaths. If the breathing rate is twenty complete breaths per minute, estimate the rate at which the work of breathing is done.

It is further estimated that the breathing muscle efficiency is about 5 per cent. Hence, calculate the rate of energy expenditure necessary to maintain breathing. If the basal metabolic rate is about 85 W, what percentage does the work of breathing contribute to the total basal rate?

What would the percentage contribution be for the heart, if:
(a) 1 J of work is done during each heartbeat;
(b) the time for one heartbeat is about 1 second;
(c) the efficiency of the heart is 15 per cent near the resting state?

4 Discuss *briefly* the major factors affecting the body's power consumption.

A hiker of mass 70 kg is found to expend power at about three times his basal metabolic rate of 80 W when walking on a horizontal track. During the course of a two-hour walk, he is found to have ascended 600 m. Assuming his body has an overall mechanical efficiency of 15 per cent, estimate:
(a) the average mechanical power necessary to overcome the altitude difference;
(b) the average rate at which his body uses energy in overcoming this altitude difference;
(c) his total average rate of energy expenditure during the walk;
(d) his rate of heat production;
(e) the number of buns he must eat at the end of the walk to replace his lost energy, given that each bun has an energy content of 10^6 J.
State any assumptions you make.

5 Describe how you would investigate whether the law of conservation of energy holds true for the human body. Include in your answer any precautions you would take and what results you would expect.

6 It is found that the average quantity of energy available by 'burning' 1 kg of fat in the body is 37.7×10^6 J. A man wishes to loose 3 kg of his body fat by playing football. Assuming that all his extra activity is at the expense of his store of fat, how many games of football must he play to loose his 3 kg? You may assume that his average rate of energy expenditure during each $1\frac{1}{2}$-hour game of football is 500 W.

Do you consider that exercise is the best way to loose weight? Discuss briefly.

2 | Human mechanics

Muscles

Skeletal muscle is attached to a bone via a strong inelastic cord, known as a tendon. When a muscle contracts, its fibres get shorter and fatter and so pull the two parts of the body together. When the muscle relaxes, it merely stops pulling, but it cannot actively expand and push the body segments apart. Hence, to return to the original state, a second muscle is needed acting in opposition or antagonism. Such antagonistic pairs of muscles are usually found acting on either side of a joint.

Human levers

The muscular–skeletal system of the body is a complex system of levers. The 'effort' provided by muscular contraction is applied to the skeleton at one point in order to move a 'load' elsewhere. The load may simply be the weight of the particular body segment being moved, or it may include an external load. The load is often further from the fulcrum (joint) than is the effort, thus giving the system a mechanical 'advantage' (MA) of less than one since:

$$\text{MA} = \frac{\text{load } (L)}{\text{effort } (E)} = \frac{\text{distance of } E \text{ from fulcrum}}{\text{distance of } L \text{ from fulcrum}}$$

This mechanical 'disadvantage' of certain muscles has its compensation in the fact that the distance moved by the effort is consequently much smaller than that moved by the load. Hence, wide arcs of movement can be achieved at speed, but at the expense of large muscle forces.

Friction at the joints is minimised by the presence of a lubricant called synovial fluid, and hence the body's lever systems waste little energy and yield high efficiencies.

Most of the body's levers are third-class levers, although examples of all three classes can be found (see Fig. 2.1).

Free body analysis

This is a technique for investigating parts of the body rather than the entire system. The part in question is treated as a free body, as long as the appropriate interacting forces are indicated across the boundary with the rest of the body.

For example, consider the free body analysis of the forearm when there is a weight in the hand, (Fig. 2.2). The boundary is taken through the elbow joint, so that the reaction R of the humerus on the forearm bones must be included,

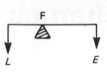

MA may be > 1 or < 1

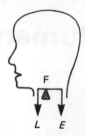

L = weight of head
E = contraction of neck muscles
F = atlas vertebra

(a) First-class levers

MA > 1
$E < L$

L = weight of body
E = contraction of calf muscles
F = ball of foot

(b) Second-class levers

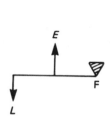

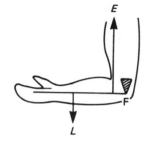

MA < 1
$E > L$

L = weight of hand and forearm
E = contraction of biceps muscle
F = elbow joint

(c) Third-class levers

Figure 2.1 Human levers

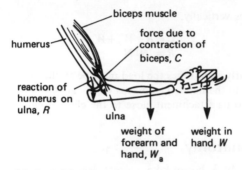

(a) Actual situation

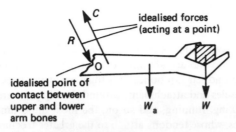

(b) Idealised free body
analysis of situation (a)

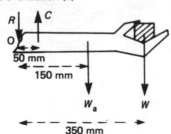

(c) Elbow flexed at 90°

Figure 2.2 Free body analysis of the forearm

whereas the equal and opposite reaction on the humerus is omitted since it does not act on the free body. In the simple case when the elbow is flexed to a right angle with the forearm horizontal, C (the contractile force of the biceps muscle) and R act vertically, (Fig. 2.2(c)). Using the data on the figure, and assuming that the weight of the forearm and hand, W_a, is 20 N and the weight held in the hand, W, is 120 N, C and R may be calculated as follows:

Taking moments about O,

$$C \times 50 = W_a \times 150 + W \times 350$$
$$C \times 50 = (20 \times 150) + (120 \times 350)$$
$$\therefore \qquad C = 900 \text{ N}$$

Resolving the forces vertically,

$$C = R + W_a + W$$
$$\therefore \qquad R = 760 \text{ N}$$

Thus, a force of nearly eight times the load must be applied by the biceps muscle to balance the weight in the hand. The biceps works at this considerable mechanical disadvantage due to its attachment close to the elbow joint.

Elementary anatomy

The hip and lower back bones take a great deal of strain in so many physical activities that it is important to have some understanding of their basic structure.

Hip joint

The femur, or thigh bone, extends from the knee to the hip, and at the upper end its curved head fits into the acetabulum to form a ball-and-socket hip joint (see Fig. (2.3)). The right and left hip bones join together at the back with the sacrum and coccyx. The muscles and attachments around the hip joint are complex but for the purposes of walking, running, and so on, the most important muscles are the hip abductor muscles, whose tendons attach to the greater trochanter.

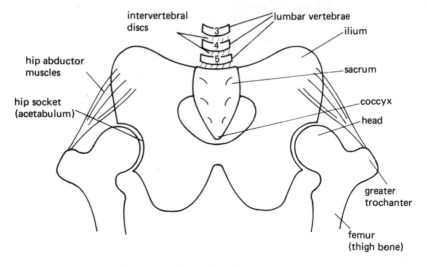

Figure 2.3   The pelvic girdle, front view

Lower vertebral column

The sacrum is rigidly attached to the pelvis and hence can only 'tilt' along with the pelvis (see Fig. 2.4). Directly above the sacrum and its curved 'tail', the coccyx, rest five lumbar vertebrae. These are separated from each other and the sacrum by intervertebral discs the lowest of which is known as the lumbosacral disc. Each disc is a self-contained fluid system which absorbs shock, uniformly transmits

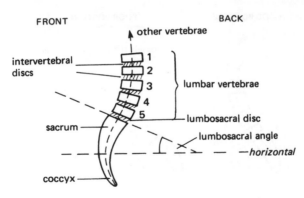

Figure 2.4 The lower vertebral column, side view

pressure, and permits deformation of the intervertebral spacing, thus allowing motion.

The angle at which the pelvis is held by its associated complex of muscles determines the curve of the line of the lumbar vertebrae. The angle between the horizontal and the top surface of the sacrum is called the lumbosacral angle, which is normally about 40°, and the resulting curve of the line of the lumbar vertebrae is called the lumbar lordosis.

Not only do the intervertebral discs have to withstand prolonged and considerable stresses, but they also suffer some degeneration with age. As a result, lower back pain is a common complaint affecting about 80 per cent of the population, and more severe ailments like muscle spasm and slipped discs are not unusual.

Standing erect on two legs

Forces at the lumbosacral joint

In the upright posture, the body's centre of gravity (CG) is located within the pelvis. In order to maintain balance when standing, this CG must lie vertically over a point within the base formed by the feet. The resulting stance may exhibit good or poor posture, (see Fig. 2.5), the latter often leading to low back pain.

Figure 2.6 shows the forces acting on the lumbosacral disc while standing. The weight of the body above the disc, known as the superincumbent weight, is approximately $0.6\,W$, where W is the total body weight.

Assuming that the spine muscles are completely relaxed, the reaction R of the sacrum on the disc must be equal to $0.6\,W$. Resolving this reaction into components, it can be seen that:

$$\text{compressive force on disc} = R\cos\theta$$
$$\text{shear force on disc} = R\sin\theta$$

For the normal lumbar lordosis, $\theta \approx 40°$.

$$\therefore \quad \text{compressive force} = 0.6\,W\cos 40°$$
$$= 0.46\,W$$

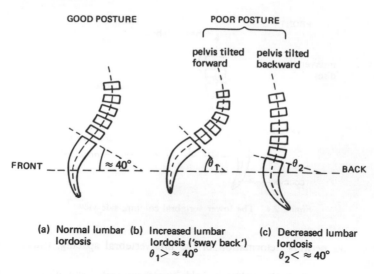

GOOD POSTURE   POOR POSTURE

pelvis tilted   pelvis tilted
forward   backward

FRONT _____≈ 40° _____θ₁_____θ₂____ BACK

(a) Normal lumbar (b) Increased lumbar   (c) Decreased lumbar
    lordosis   lordosis ('sway back')   lordosis
                       $\theta_1 > \approx 40°$ $\theta_2 < \approx 40°$

Figure 2.5   Standing erect, side view

and

$$\text{shear force} = 0.6\,W \sin 40°$$
$$= 0.39\,W$$

This is a considerable shear force on a disc which, by its structure, is less capable of withstanding shear than compression. If, in addition, there is 'sway back' as caused, for instance, by weak abdominal muscles, weak hip flexor muscles, or pregnancy, the shear force increases in proportion with $\sin \theta$. This produces a tendency for the lower lumbar vertebrae to slip forward, causing irritation and pain. The obvious course for the relief of such pain involves exercises to flatten the lumbar lordosis.

Forces at the hip

In the erect stationary position, the body is equally balanced on both legs and the muscles around the hips are relatively inactive. The load F on each femoral head must then (like the reaction R of the femoral head on the acetabulum) be vertical and equal to half the superincumbent weight, which here is about $0.7\,W$. Thus:

$$F = R = 0.5 \times 0.7\,W = 0.35\,W$$

Since the femoral neck generally forms an angle of about 50° with the vertical (Fig. 2.7), this force may be resolved into shear and compressive components:

$$F_{\text{shear}} = F \sin 50° = 0.35\,W \sin 50°$$
$$= 0.27\,W$$
$$F_{\text{compressive}} = F \cos 50° = 0.35\,W \cos 50°$$
$$= 0.22\,W$$

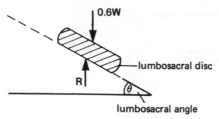

(a) Forces acting on the disc

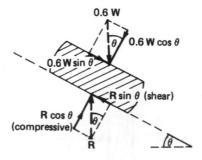

(b) Compressive and shear components of the forces

Figure 2.6 Free body diagram of the lumbosacral disc

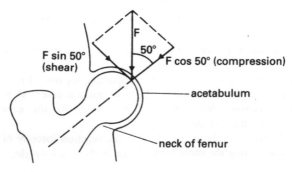

Figure 2.7 Force at the hip while standing on two legs

Standing on one leg

Static situation

When one foot is lifted from the ground, in order to maintain balance the body's CG must lie vertically above the supporting foot. The superincumbent weight then no longer acts through the femoral head, and the resulting moment it produces about the femoral head tends to make the pelvis drop on the unsupported side.

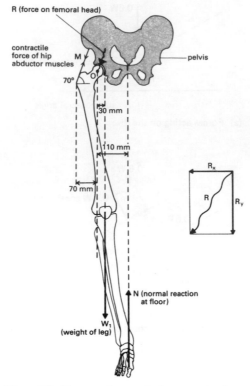

Figure 2.8   Forces acting on single supporting leg

This drop is counteracted by contraction of the hip abductor muscles on the stance side.

Figure 2.8 shows the forces acting on the supporting leg only. From X-ray films, the line of action of the effective hip abductor muscle force M acting at the position of the greater trochanter, is found to lie at about 70° above the horizontal. O is the centre of the head of the femur.

Considering the whole body in equilibrium, the reaction force N of the ground on the foot must act along the same vertical line as the body weight, and be equal to it; thus:

$$N = W$$

The weight of the leg W_1 is approximately equal to 0.15 W. Taking moments about O:

$$N \times 110 = W_1 \times 30 + M \sin 70° \times 70$$
$$\therefore \qquad 110\,W = 4.5\,W + 65.8\,M$$
$$\therefore \qquad M = 1.6\,W$$

Thus, the hip abductor muscle force is more than $1\frac{1}{2}$ times the total body weight.

The horizontal (R_x) and vertical (R_y) components of the force R on the femoral head may be found by balancing horizontal and vertical forces.
Horizontally:

$$R_x = M \cos 70° = 0.55\,W$$

Vertically:

$$R_y + W_1 = M \sin 70° + N$$
$$\therefore \qquad R_y = 2.35\,W$$

Then

$$R^2 = R_x^2 + R_y^2$$
$$\therefore \qquad R = 2.42\,W$$

The head of the femur is thus subjected to a considerable force of approximately $2\frac{1}{2}$ times the body weight. This is about seven times the loading it receives when the body is supported on two legs ($R \approx 0.35\,W$).

Dynamic situation

The hip forces experienced during walking are similar to those calculated above for the body supported on one leg. Once again R assumes a high value mainly due to the large moment of N about O. Any change in the positioning of the foot relative to the hip can significantly change this moment and hence affect the forces M and R. Two examples are given below.

(a) Using a cane

The use of a cane when walking broadens the base above which the body's CG must lie. It is thus possible for the foot on the opposite side to the cane to be placed more directly underneath its hip during the support phase of the stride. This reduces the moment of N (itself reduced) about O and consequently reduces M and R. Thus, following hip surgery, using a cane on the side opposite to the weak hip is recommended.

(b) Carrying an object

Carrying a case, for example, when walking requires the body to tilt away from the case to maintain balance. The hip on the side opposite the case thus experiences larger forces due to the increased moment of N (itself increased) about O.

Bending and lifting

Bending

When bending the back or lifting objects from low positions, the muscles most involved are the erector spinae or sacrospinal muscles. These link the lower sacrum and ilium with each of the lumbar and four of the thoracic vertebrae. The situation may be simplified by regarding the vertebral column as a rigid body, rotating about a fixed fulcrum, namely the lumbosacral joint (see Fig. 2.9). The superincumbent weight of the head, arms and trunk together, W_s, acts through their combined CG at C, which is about two-thirds of the way up the trunk. X-ray measurements indicate that the resultant contractive force of the erector spinae muscles, E, then acts on the vertebral column at C and is inclined to it at an angle of about 10°.

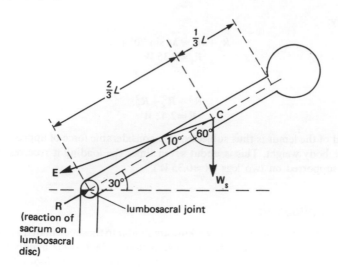

Figure 2.9   Forces on the vertebral column during bending

Consider bending through an angle of 60° from the vertical, with the arms hanging freely (see Fig. 2.9). Since E and W_s are assumed to act through the same point C in this simplified treatment, the line of action of R must also pass through C to maintain equilibrium.

$$W_s \approx 0.6\, W = 500\,\text{N} \quad \text{(for an average man)}.$$

Resolving forces at right angles to the vertebral column gives:

$$E \sin 10° = W_s \sin 60°$$
$$\therefore \qquad E \approx 3\,W \quad \text{(about 2500 N)}$$

Resolving forces along the vertebral column gives:

$$R = E \cos 10° + W_s \cos 60°$$
$$\therefore \qquad R \approx 3.25\,W \quad \text{(about 2700 N)}$$

Thus R is more than three times the total body weight (a value about seven times greater than that experienced during normal standing) and produces a contraction of the lumbosacral disc of about 20 per cent!

Lifting

When a weight (e.g. 200 N) is held in the hands during bending (Fig. 2.10), there are no longer three forces in equilibrium, and R does not act along the vertebral column but at a small angle, θ, to it. If R_c is its compressive component, and R_s the shear component, taking moments about O gives:

$$E \times \tfrac{2}{3} L \sin 10° = W_s \times \tfrac{2}{3} L \sin 60° + 200 \times L \sin 60°$$

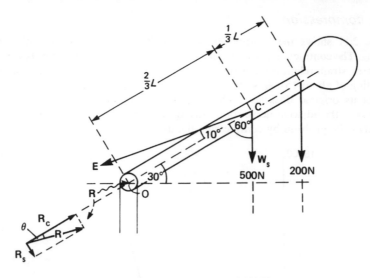

Figure 2.10   Lifting a load of 200 N

Substituting $W_s = 500$ N   gives:

$$E = 3990 \text{ N}$$

Resolving forces along the vertebral column:

$$R_c = E \cos 10° + W_s \cos 60° + 200 \cos 60°$$

∴ $$R_c = 3990 \cos 10° + (500 \times \tfrac{1}{2}) + (200 \times \tfrac{1}{2})$$

∴ $$R_c = 4280 \text{ N}$$

Resolving forces at right angles to the vertebral column:

$$E \sin 10° = W_s \sin 60° + 200 \sin 60° + R_s$$

∴ $$R_s = 86.6 \text{ N}$$

The total reaction R is then found using:

$$R^2 = R_c^2 + R_s^2$$

∴ $$R = 4281 \text{ N}$$

and R acts at an angle θ to the vertebral column, where:

$$\tan \theta = \frac{R_s}{R_c}$$

∴ $$\theta \approx 1°$$

The compressive force R_c on the lumbosacral disc in this case is seen to be considerable, being more than eleven times the size of the compressive force produced in normal standing, and more than five times the total body weight.

Disc compression

Figure 2.11 shows the compression of a healthy lumbar vertebral disc under loading. The contraction is elastic up to a loading of about 1000 N but beyond this the stress–strain relationship is markedly non-linear. Rupture of the disc occurs at a loading of about 15 000 N, at which point the disc contraction is about 35 per cent of its original thickness. When rupture occurs, extrusion of the disc fluid compresses the adjacent nerve root, causing pain and muscle spasm, which in turn magnifies the problem by compressing the disc even further.

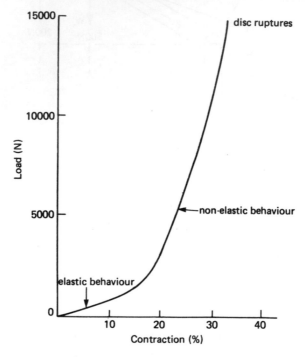

Figure 2.11   Behaviour under loading of lumbar vertebral disc

Ground forces

Reaction

When standing still on two legs, with the body weight distributed evenly, the normal ground reaction R at each foot is approximately equal to half the body weight. The pressure exerted on the ground, however, is inversely proportional to the area of contact, and Fig. 2.12 illustrates how this can vary considerably. Not only can small contact areas be damaging to the supporting surface (e.g. stiletto heels), but also to the supported body (e.g. bed sores).

When walking, the body weight is shared, generally unequally, between the two feet. For simplicity, it is assumed that only one foot is in contact with the ground

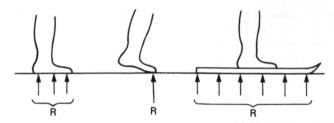

(a) Whole foot in contact   (b) Toe contact only: (c) Ski or snowshoe worn:
                                       small area, high   large area, low pressure
                                       pressure

Figure 2.12   Varying pressure for the same ground reaction

at a given moment, going through the processes of heel-strike, support and toe-off (Fig. 2.13), while the other is swinging through the air. Although the normal ground reaction R_2 during the support stage is approximately equal to the body weight W, the reaction during the other phases (R_1 and R_3) both exceed W due to the existence of other vertical forces:

(a) at heel-strike, an additional force is needed to destroy the leg's downward momentum;
(b) at toe-off, an extra force is required to produce some upward acceleration. This is the reaction to the thrust of the calf muscles.

The average normal ground reaction thus increases as the body progresses from standing to walking to running, when even greater additional vertical forces are necessary to sustain the 'leaping' action.

Friction

Frictional forces are necessary during walking, running, and so on, to prevent the foot from slipping at heel-strike (F_1) and to provide the forward thrust (F_2) at toe-off, (Fig. 2.13). For a given normal reaction R, the maximum frictional force, known as limiting friction, is μR, where μ is the coefficient of static friction between the two surfaces. Hence, if there is to be no slipping, frictional force $\leqslant \mu R$.

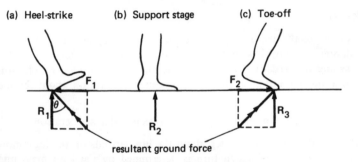

(a) Heel-strike   (b) Support stage   (c) Toe-off

Figure 2.13   Ground forces during walking

The larger values of R at heel-strike and toe-off increase the value of limiting friction and hence tend to stabilise the foot. For an average walking surface, F_1 and F_2 are approximately 15 per cent and 20 per cent of the body weight respectively, and slip does not occur.

Resultant ground force

The resultant ground force, G (Fig. 2.14) may be found using:

$$G = \sqrt{(R^2 + F^2)} \qquad \text{(magnitude)}$$

$$\tan \theta = \frac{F}{R} \qquad \text{(direction)}$$

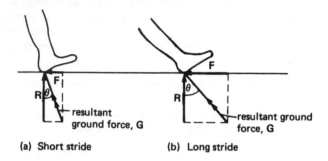

(a) Short stride   (b) Long stride

Figure 2.14   Effect of stride length on slipping

The condition for no slipping is $F \leqslant \mu R$, that is:

$$\tan \theta = \frac{F}{R} \leqslant \mu$$

Thus, slipping is more likely to occur for (i) small values of μ (icy, slippery surfaces), and (ii) large values of θ (long strides).

Walking

The positions of the leg during the normal walking cycle (Fig. 2.15) have been analysed, mainly using slow-motion film and stroboscopy. After toe-off, the whole limb is swung forward by the action of the hip flexor muscles, F. The torque so produced about the hip gives the leg an angular acceleration about the hip given by

Torque = moment of inertia (MI) × angular acceleration

Thus, the initial rotation of the leg about the hip depends on the leg's moment of inertia about the hip, which in turn is determined by the leg's mass and mass distribution. If the normal mass distribution should change for any reason (for

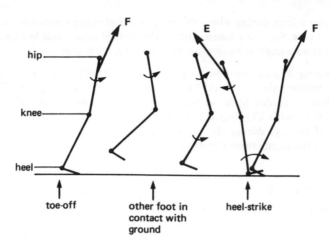

Figure 2.15   Steady walking

example after amputation or the attachment of a brace or artificial limb, or when wearing heavy boots), the torque required at the hip also changes and can easily lead to excessive energy expenditure and even muscle fatigue.

After the initial acceleration, the leg swing continues with constant angular velocity without much further effort, in accordance with Newton's first law of motion, until the swinging limb is stopped, or decelerated, in a controlled fashion by the hip extensor muscles, E. Gravity then further assists in bringing the heel to the ground.

In addition to the torque about the hip, there is also a torque about the knee joint, rotating the lower limb about the knee. This brings the foot in front of the body as the body's CG moves forward. The forces which produce this torque arise from the ligaments and muscles around the knee joint, gravitational forces and joint reactions.

The main effort required in walking in simply that of swinging the legs, and if these are allowed to oscillate at a natural rate the process absorbs little energy. It is much more tiring to walk at an unnatural pace or to keep changing speed as, for example, when walking with children. Changes in speed are best made by changing the length of the stride, rather than changing the natural frequency of the leg oscillations.

Running

Running speed

A good sprint speed of 9 ms^{-1} is six times faster than the average walking speed of 1.5 ms^{-1}. This factor of six is achieved by:

(a) increasing the number of strides per second by a factor of about two; and
(b) increasing the stride length by a factor of about three. This is accomplished

by using a leap, during which the body is out of contact with the ground for a short time. Since any momentum in the vertical direction is lost on landing, the optimum angle of projection for the leap is fairly low.

When running at a constant velocity, little muscular effort goes towards maintaining the velocity since even at $9 \, \text{ms}^{-1}$, air resistance is not great. The main muscular drive is needed to move the legs sufficiently quickly, that they remain underneath the body's CG. Indeed, the major factors limiting running speed are the inertia of the legs during the 'leap' and the forces that must be applied to accelerate and decelerate them during this phase.

The running cycle

The leg movement during a running cycle, (Fig. 2.16) consists largely of a combination of rotations about the hip and knee joints. Just before toe-off, the supporting leg, which is kept fairly straight when in contact with the ground, is

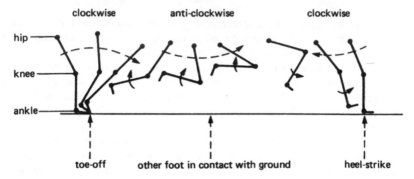

Figure 2.16   Leg movements during one complete running cycle of two strides

rotating in a clockwise direction, possessing clockwise angular momentum. Soon after toe-off, the thigh starts rotating in an anticlockwise direction in order to carry the upper leg forward. This is brought about by:

(a) an anticlockwise torque about the hip

anticlockwise torque $= \text{M.I.} \times$ anticlockwise angular acceleration

(b) a clockwise torque about the knee which,
    (i) bends the leg about the knee, thereby reducing the leg's MI about the hip, and increasing the anticlockwise angular acceleration of the thigh in (a);
    (ii) increases the clockwise momentum of the lower leg, and so (by the principle of conservation of angular momentum applied to the whole leg) transfers an equivalent anticlockwise momentum to the thigh.

When the thigh has rotated into an almost horizontal position, the above torques are reversed, angling the leg down in preparation for heel-strike.

The body as a whole twists about its own axis during running, although the upper and lower bodies twist in opposite directions in order to conserve total angular momentum. Thus, at the instant shown in Fig. 2.17, the anticlockwise angular momentum of the upper body, generated by the action of the abdominal and upper-body muscles, results in an equal and opposite clockwise angular momentum being transferred to the lower body, enabling the left leg to swing forward, as shown.

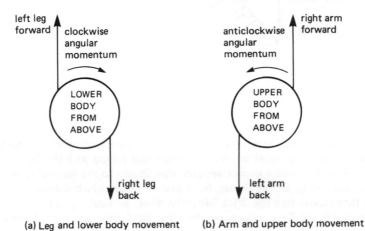

(a) Leg and lower body movement   (b) Arm and upper body movement

Figure 2.17   Total body movement during running

Energy conversions in jumping and falling

During jumping, muscular work appears as kinetic energy $(KE = \frac{1}{2}mV^2)$ at take-off. This in turn is converted into potential energy, $(PE = mgh)$, which allows the jumper's CG to be raised through a maximum height

$$h_{max} = \frac{V^2}{2g}$$

Certain techniques may increase the height achievable in, for example, the high jump: swinging the 'free' limbs upwards to generate upward momentum and 'diving' over the bar to keep the body's CG low, can both improve performance.

Falling involves the reverse energy exchange, namely the conversion of PE into KE. If the CG falls a distance x, the velocity V on impact is given by:

$$V = \sqrt{(2gx)}$$

In the simple case of a constant decelerating force F acting for a time t to bring a body to rest during impact:

$$F = \frac{m\sqrt{(2gx)}}{t}$$

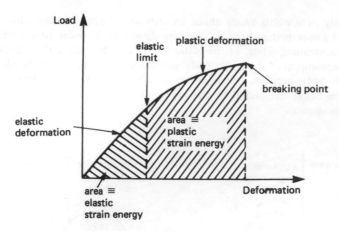

Figure 2.18   Strain energy

Thus, to reduce F, and hence the likelihood of fractures occurring, t should be increased by such actions as bending the knees and rolling with the fall.

Much of the KE before impact appears after impact in the form of elastic and plastic strain energy in soft tissues, bone and cartilage, which deform under the loading they receive (see Fig. 2.18). Muscular work, particularly in the limbs, and strain energy in the floor surface, are the other chief areas of energy dissipation.

Exercise 2

(Take $g = 10\,\mathrm{ms}^{-2}$ throughout)

1 Describe how the body uses the principle of levers, including in your answer at least one example of the three classes of levers.

   When a mass of 11 kg is held in the hand and the elbow flexed at 90° so that the contractile force of the biceps acts vertically upwards, the lines of action of the muscle force, combined weight of hand and forearm, and the weight in the hand are found to act at distances of 5 cm, 15 cm and 30 cm respectively from the idealised point of contact between the lower and upper arm. If the hand and forearm together have a mass of 2 kg, find:
   (a) the force provided by the biceps;
   (b) the reaction between the upper and lower arm bones.

2 Explain why bad posture can easily lead to pain in the lower back.

   A woman of mass 50 kg stands erect with a lumbosacral angle of 40°. Find the shear and compressive forces acting on her lumbosacral disc if the superincumbent weight here is 60 per cent of the body weight.

   Consider now the same woman eight months pregnant. Find the new shear and compressive forces acting on the lumbosacral disc if the superincumbent weight is increased by 50 N and the lumbosacral angle increased by 10°.

3 If the weight of the trunk is 40 per cent of the body weight and acts half-way along the vertebral column, and the weight of the head and arms is 20 per cent

of the body weight and acts at the top end of the vertebral column, find the position of the combined centre of gravity of the trunk, head and arms.

Assuming the erector spinae muscles act through this combined CG at an angle of 10° to the vertebral column, find the erector spinae muscle force when:
(a) a 60 kg woman bends forward through an angle of 30° to the vertical, keeping her arms hanging vertically downwards;
(b) the same woman in the same position carries a load of 300 N in her hands.

Why is it better to bend the knees and keep the back straight when lifting a load, such as a case, from the ground?

4 Explain how the conservation of angular momentum is important in the process of running.

What factors limit the speed of a sprinter? Why does he use a starting block?

A runner of mass 70 kg accelerates from $6\,\mathrm{ms}^{-1}$ to $8.5\,\mathrm{ms}^{-1}$ in 4 s. Assuming uniform acceleration, and neglecting air resistance, calculate:
(a) the average frictional force between his feet and the ground;
(b) the distance he travels in this time.

Would multiplying your two answers together give you his energy expenditure? Discuss.

5 Describe the energy changes that occur as an athlete performs the high jump. How are these modified if he uses a pole as in the pole vault?

In a vertical jump, a man of mass 70 kg crouches down, then jumps upward by straightening both legs and throwing his arms upward. The time taken from the bottom of the crouch to take-off is 0.25 s and the depth of the crouch is 0.4 m. Assuming his centre of gravity undergoes uniform acceleration during the spring, calculate:
(a) the take-off velocity;
(b) the height to which his centre of gravity will go, given that at take-off it is 1.1 m above the ground;
(c) the mechanical work produced by the muscles for one jump;
(d) the average power developed by the muscles to produce the jump.

6 Using free body analysis, discuss the forces acting on the supporting leg when standing on one leg.

Explain why:
(a) the pain due to an arthritic right hip increases as the patient starts walking slowly;
(b) the use of a cane held in the *left* hand helps to reduce this pain.

7 Explain the following:
(a) Hobnails on boots are useful in icy weather.
(b) It is hard work to walk on soft sand.
(c) Shorter strides on ice are less hazardous than longer strides.
(d) When a person disembarks from a canoe, the boat tends to move away from the jetty.

The leg of a man of mass 100 kg first makes an angle of 20° with the vertical during heel-strike. Neglecting the downward momentum of the foot, and assuming the resultant ground force acts along the direction of the leg, estimate:

(e) the total ground force G;

(f) the frictional force F between his foot and the ground. (You may assume his other foot is not in contact with the ground, and that limiting friction is not exceeded.)

If the man increases his stride length, so that during heel-strike the angle his leg first makes with the vertical is $30°$, calculate the new G and F, and show that the percentage increase in F is greater than the percentage increase in G.

8 With particular reference to the laws of motion and conservation of energy, describe the action of a diver climbing to a high spring-board, running and diving from the board.

(a) Why might a 'belly-flop' be injurious to the diver?

(b) Why does a diver curl to increase his spinning capability?

An upright man of mass 70 kg jumps from a wall 2 m high on to a solid floor. On impact with the ground, he bends his knees, so that he decelerates steadily from his landing speed to zero while his centre of gravity moves a further distance of 0.4 m. Calculate:

(c) his initial landing speed;

(d) the time of duration of the decelerating crouch;

(e) the average reaction force from the ground during the impact;

(f) the stress on each tibia (lower leg bone) during impact. (The average cross-sectional area of the tibia is $3.5 \times 10^{-4} \, m^2$.)

If the compressive breaking stress for bone is $1.5 \times 10^8 \, Nm^{-2}$, will the jumper suffer a fractured tibia?

If he performed the same jump but kept his legs stiff so that during impact his centre of gravity moved a further distance of only 10 mm, how would the situation change?

3 | The eye

Structure of the eye

The eye (Fig. 3.1) is filled with fluid to maintain the even spherical shape of the eye wall and is divided into two chambers by the lens. This transparent, elastic biconvex lens is not homogeneous, but consists of successive fibrous layers gradually added throughout life. It has a refractive index which varies from about 1.42 at the centre to about 1.38 at the periphery.

The light-sensitive part of the eye is the layered retina, shown in more detail in Fig. 3.2. The light-sensitive receptors are known as rods (slender rod-like elements) and cones (narrow conical elements), and have diameters of a few microns[1]. There are about twenty times as many rods as cones, with rods dominating the retinal periphery and cones the optical axis.

Two well-defined spots may be observed on the retina:

(a) The blind spot

This creamy white spot, situated about 10° from the optical axis on the nasal side, is where the optic nerve leaves the eye. Since there are no receptors at this spot, it

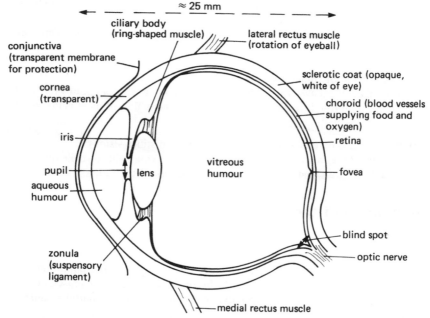

Figure 3.1   Horizontal section of right eye

[1] 1 micron = 1 μm = 10^{-6} m.

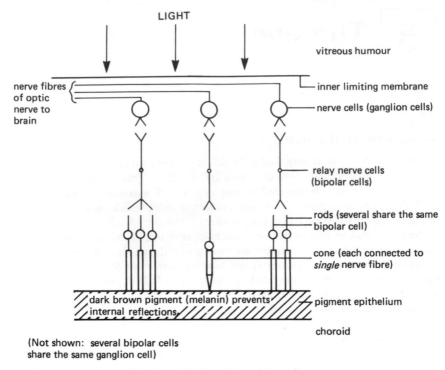

Figure 3.2   Structure of the retina

is responsible for a 'blind field', ($\approx 6°$ horizontally, $8°$ vertically) when using one eye only.

(b) The yellow spot (macula)

Situated on the optical axis, this yellowish spot has a diameter of about 1 mm with a small depression (the fovea) at its centre. This is the region of most acute vision since:

(i) it contains only cones (the receptors of most distinct vision);
(ii) the nerve fibres are displaced radially away, allowing light to strike the cones directly without first traversing the other layers of the retina.

Functioning of the eye

Brightness control

The iris controls the amount of light entering the eye through the pupil and thus constitutes the eye's brightness control. In bright light, the circular muscles of the iris contract thereby constricting the pupil; in dim light the radial muscles contract dilating the pupil.

The pupil in fact oscillates about a mean size in conditions of constant illumination due to a negative feed-back mechanism. The iris opens and more light reaches the retina which then signals back to the iris to close. The iris closes and less light reaches the retina, which then signals the iris to open, and so on.

Convergence

Since the foveal region of the retina is the area of most acute vision, the eye muscles always orientate the eyeballs so that their visual axes are directed towards the object studied. If the object moves towards the eyes, the eye muscles automatically pivot the eyes to 'follow' it, so that the visual axes converge.

Accommodation

In order to focus rays from objects at various distances from the eye, the eye lens can change its shape and hence its refractive power. This automatic mechanism is known as accommodation.

The exact shape of the lens is determined by the seventy or so suspensory ligaments (zonula) attached radially around the lens, pulling its edges towards the ciliary body. When the eye is accommodated for distant vision, both the circular and meridional fibres of the ciliary muscle are relaxed, thus stretching the zonula which squeezes the elastic lens to a flattened shape (see Fig. 3.3). In order to

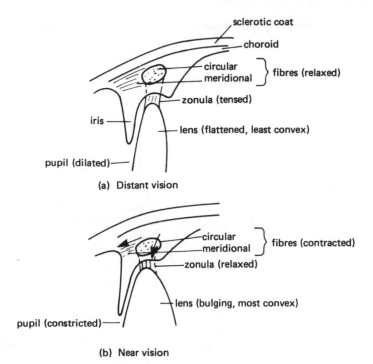

Figure 3.3   Accommodation

increase the refractive power of the lens when viewing a close object, both sets of fibres of the ciliary muscle contract. Each has the effect of releasing the tension in the zonula, which then allows the lens to bulge.

The optical system of the eye

Refractive power

The power of a refracting surface is defined as:

$$\text{power (in dioptres, D)} = \frac{1}{\text{focal length (in m)}}$$

The powers of successive refracting surfaces may be added algebraically, an advantage over using focal length.

The power of the unaccommodated eye is approximately 59 D, refraction occurring at several interfaces (see Fig. 3.4). Most of the refraction (≈ 46 D) occurs at the air/cornea boundary due to the large change in refractive index across this interface, and this amount remains constant regardless of object distance. The lens provides only about 18 D and the increase in power of the lens for near vision is a fine adjustment of the focusing system. The total range of lens power during

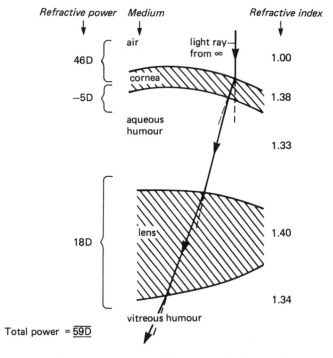

Figure 3.4   Refracting surfaces of the unaccommodated eye and their powers

accommodation is known as the amplitude of accommodation. In young people, the lens power increases from about 18 D for distant vision to about 29 D for near vision, giving an amplitude of accommodation of about 11 D. In young children, the amplitude of accommodation may reach 14 D, and in elderly people, it approaches zero.

In a very simple model of the unaccommodated eye, all refraction is assumed to take place at a single equivalent eye lens of power 59 D.

Far and near points of the eye

The far point of the eye is the most distant point it can focus, and for the normal eye it is infinity. The closest point at which an object may be seen clearly is known as the near point, and for the normal eye this least distance of distinct vision is about 250 mm.

Depth of focus

When two point sources of light are observed by an eye with a small pupil (see Fig. 3.5(a)) the retinal spot size, and hence clarity, does not change much over quite a

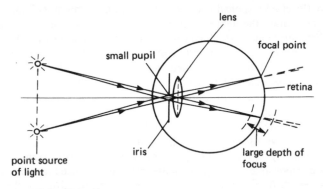

(a) Small pupil

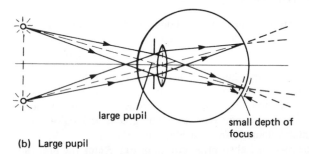

(b) Large pupil

Figure 3.5 Depth of focus

wide range of image positions. There is thus said to be a large depth of focus. In contrast, a large pupil (see Fig. 3.5(b)) produces a small depth of focus, whereby even a small change of focus from the retina produces enlargement and consequent blurring of the spots on the retina.

A large depth of focus is particularly desirable when viewing near objects, when relatively small changes in object distance can lead to large changes in image position. Hence, the pupils constrict, not only in bright light, but also when observing close objects.

The response system of the eye

The inverted optical image formed on the retina causes light-sensitive chemicals in the rods and cones to decompose, thus stimulating electrical impulses in the associated nerve fibres leading to the brain. The rods and cones display different characteristics:

(a) Rods do not register colour; they operate at low light intensities and share neural pathways to the brain.
(b) Cones discriminate colour, are stimulated only by fairly high light intensities and have less sharing of nerve fibres, particularly near the foveal region.

Rods are therefore responsible for vision in dim light (night, twilight or scotopic vision). They contain the light-sensitive pigment rhodopsin, (or visual purple due to its reddish-purple hue), which consists of a protein (scotopsin) and a type of the pigment retinene which is chemically similar to vitamin A. On exposure to light, rhodopsin is bleached, first yellow then clear, thereby stimulating the rod cell and its associated nerve cells. Bleached rhodopsin is rapidly restored to its original condition by an enzymatic process requiring vitamin A derived from the blood supply in the choroid.

Although the rods do not supply any information concerning colour, they are not equally sensitive to all wavelengths, displaying maximum absorption in the green (500–580 nm) region of the spectrum, at about 510 nm (Fig. 3.6(a)).

Cones are responsible for the more acute vision experienced in ordinary daylight conditions (colour or photopic vision). Between them, they discern all wavelengths from about 400 nm to 750 nm, producing the greatest sensitivity (in the green) at around 555 nm, (Fig. 3.6(a)).

It is now generally accepted that there are three types of cone in the retina:

(a) a red-sensitive cone (or 'red cone'), containing the pigment erythrolabe;
(b) a green-sensitive cone (or 'green cone'), containing the pigment chlorolabe;
(c) a blue-sensitive cone (or 'blue cone'), containing the pigment cyanolabe.

Each pigment is thought to consist of a protein (photopsin) and a retinene and, as in the rods, a bleaching process is believed to initiate the nerve impulse.

Figure 3.6(b) shows the approximate contribution to overall cone absorption from the three groups, and illustrates why greens and yellows are normally found to be brighter colours than deep blues or reds.

Using the re-scaled absorption curves of Fig. 3.6(c), the brain's interpretation of colour may be understood. For example, monochromatic light of wavelength

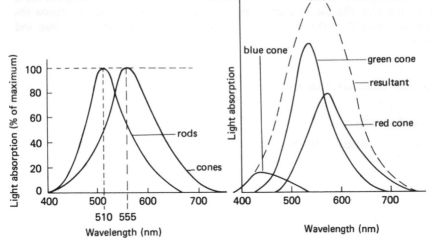

(a) Rod and cone absorption curves

(b) Relative contributions to cone absorption

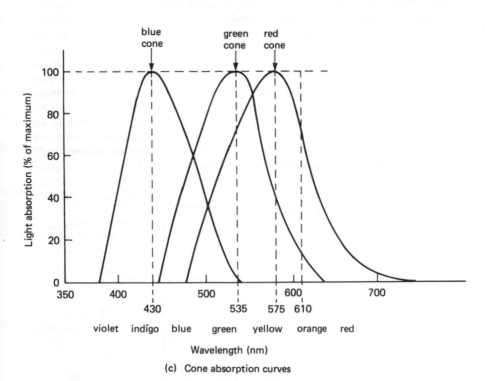

(c) Cone absorption curves

Figure 3.6   Receptor absorption curves

610 nm stimulates the red cones to a 'stimulus value' of 75 per cent, the green cones to 13 per cent and the blue cones to 0 per cent. The brain then interprets the stimulus ratios 75:13:0 as red, whilst the ratio 0:14:86 is interpreted as blue, and so on.

Resolution of the eye

Visual acuity

Visual acuity describes the fineness of detail discernible by the eye. If two points of light which are seen as just separate (or just resolved) subtend an angle of θ minutes at the eye, then the visual acuity is $1/\theta$ per minute. Several factors influence the eye's visual acuity, as detailed below.

(a) The optical system
Some blurring of retinal images always occurs due to:

(i) aberrations (see page 41) which increase with increasing pupil size;
(ii) diffraction of light mainly as it passes through the pupil. Diffraction increases as pupil size decreases.

Hence, the retinal image of a distant point source of light is a spot being bright at the centre and shading off towards blurred edges. The best compromise between aberrations and diffraction occurs with a pupil diameter of about 4 mm, and this then leads to a retinal spot diameter of about 10 μm.

When two distant point sources are observed (Fig. 3.7) the resulting retinal images may overlap. Using Rayleigh's criterion, these images are just resolved if the maximum of one image intensity curve falls on the first minimum of the other image intensity curve. Under optimum conditions, this occurs for a distance between the centres of the retinal spots of about 2 μm, and corresponds to an angular resolution of just less than half a minute. If two point sources are positioned about 10 m from the eye, about 1 mm apart they subtend approximately this angle at the eye and are just resolved by the optical system.

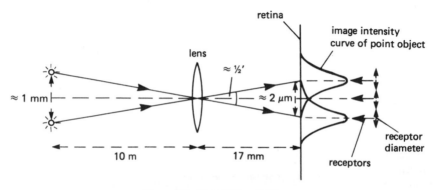

Figure 3.7   Resolution of the eye

(b) The response system

Not only must there be adequate illumination to ensure stimulation of the relevant receptors, but also the mosaic of retinal receptors must have a structure sufficiently fine to record the resolved images of the optical system.

If the separation between two adjacent retinal spots is to be perceived, there must be a receptor present in the intervening non-stimulated area. Hence the minimum perceivable separation will correspond to a spacing of at least two receptor diameters. At the fovea, the cone diameter averages about 1.5 μm and each cone has its own associated nerve fibre. Since there are only cones present here, a separation of about 3 μm can be perceived, corresponding to a visual angle of just over half a minute.

Thus, at the fovea the resolving power of the optical system approximately matches that of the receptor system providing an angular limit of resolution of about half a minute, and therefore a visual acuity of about 2 per minute.

As the image moves away from the fovea, the visual acuity for photopic vision decreases rapidly. This is due partly to the increased aberrations associated with oblique refraction, but mainly due to the increased sharing of nerve fibres by the receptors. There are approximately 6 million cones and 125 million rods in the human retina, and these have to share about a million nerve fibres of the optic nerve. More sharing is done by the rods and also more sharing occurs at the retinal periphery. Hence, the allocation of nerve fibres goes from one per cone at the fovea to about one per 600 rods at the periphery. The photopic visual acuity, therefore, varies from about 2 per minute at the fovea to about 1 per minute just outside the fovea to about 0.05 per minute at the retinal border. Standard photopic visual acuity is taken to be about 1 per minute.

In contrast to this, for scotopic vision the fovea is inactive and the region just beyond it gives maximum visual acuity. Peripheral vision is best in very weak light due to the maximum convergence of rods onto nerve fibres. For example, when astronomers wish to detect very faint stars they 'look off' the fovea so that the image falls towards the retinal periphery. On average, scotopic visual acuity is about 0.1 per minute.

Scanning

When a point source of light is focused on the retina:

(a) some nerve fibres in this area which were previously 'silent' begin to 'fire'; these are called 'on-fibres';
(b) other fibres previously discharging become silent; these are termed 'off-fibres';
(c) other fibres rapidly adapt to the existing light and discharge briefly at the beginning and end of the light stimulus only; these are called 'on-off-fibres' and are more numerous than the other two types.

Thus, if an image is fixed on the retina, after the initial stimulus a large number of the nerve fibres become inactive, resulting in a rapid fading of the image. To prevent this, the eye executes a continuous scanning movement consisting of small oscillations at about 30–80 Hz which keeps the on-off-fibres firing at the border of the image.

The eye's response to varying illumination

Action of the pupil

The diameter of the pupil can vary between about 1.5 mm and 8 mm. This provides about a thirty fold change in area and hence in light energy entering the eye. Pupillary constriction is a temporary automatic reflex mechanism mainly to protect the retina from very intense light. As the retina adapts to the new illumination, the pupil gradually returns to its original size.

Adaptation of the retina

The sensitivity of the retina to light intensity may be estimated by measuring the smallest light intensity capable of producing the sensation of vision. The smaller this 'threshold' intensity is, the greater is the retinal sensitivity. The retina is most sensitive in dim light, and as the prevailing illumination changes, sensitivity can vary by a factor as great as 10^6.

The increasing sensitivity of the retina as it goes from a light to a dark environment is known as dark adaptation. It requires a certain period of time, as shown in Fig. 3.8(a), and exhibits two distinct phases. During the first five minutes adaptation of the cones is the most important process; then the more slowly adapting rods begin to dominate. Adaptation is largely complete in about 40 minutes, but it can continue for hours.

Light adaptation is the decrease in retinal sensitivity when the eye is subjected to brighter surroundings. It occurs much more quickly than dark adaptation and does not display the two-step change (Fig. 3.8(b)).

Retinal adaptation involves two types of change, namely neural and photochemical.

(a) Neural
As the eye dark-adapts, the retina integrates energy over:

   (i) a greater area and, therefore, a greater number of receptors;
   (ii) a greater time, in a similar way to taking a time exposure in photography.

This leads to a substantial gain in sensitivity, but occurs at the expense of visual acuity in both space and time. In other words, the perception of fine detail and changes in position are sacrificed and this makes, for example, the playing of ball games difficult in dim light.

(b) Photochemical·
When the eye is suddenly subjected to bright light a large proportion of the photochemicals in both the rods and cones decomposes. Since the regeneration process is slow, the concentration of photochemicals present in the receptors falls to a low level. 'Saturation' is then said to occur, the receptors become inactive, and the sensitivity rapidly assumes the low value characteristic of the light-adapted eye.

Conversely, when the eye is plunged into darkness the decomposition of the photochemicals ceases and the slow regeneration process becomes important. As

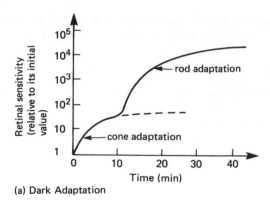

(a) Dark Adaptation

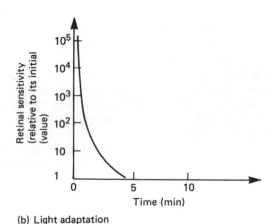

(b) Light adaptation

Figure 3.8   Retinal adaptation

the concentration of photochemicals gradually increases, so too does the retinal sensitivity. Dark adaptation is therefore a much slower process than light adaptation.

The two-step increase in sensitivity (Fig. 3.8(a)) is caused not only by the different rates at which the rods and cones resynthesise their photochemical pigments, but also by the sharing of nerve fibres by the rods, which greatly enhances their response at very low intensities.

Persistence of vision

When a light stimulus is removed, the sensation of vision takes up to about one fifth of a second to disappear, due to the persistence of excitation of the receptors. If successive flashes of light occur fast enough, 'flicker fusion' results and the light appears continuous. At low intensities, fusion occurs at rates as low as five to six flashes per second. At high intensities, (in which the brain seems to be able to detect intensity changes more easily) the critical frequency for fusion rises to about 60 Hz.

Flicker fusion is of prime importance for example in television where 60 frames per second appear on the screen, and at the cinema where films are projected at 24 frames per second. Some modern projectors have a special shutter that shows each frame three times in rapid succession, making the flicker rate 72 Hz, which is greater than the critical fusion frequency for all but the brightest lights.

Depth perception

Monocular vision

A simple trial will indicate that the single eye is capable of perceiving depth, or relative distance, to a certain extent, although exact judgements, like those necessary for aligning two pencil points so that they touch, are difficult. When only one eye is operational, the brain relies heavily on past experience to produce the sensation of depth. For example, the apparent size of a familiar object whose actual size is known, can lead to an estimation of its distance from the eye. Similarly, the eye uses the overlapping of distant objects by near ones, shadows, reflections, and the modification of colour (distant objects appear more blue) to give depth to a stationary scene.

When the head is moved, images of close objects move across the retina faster than those of distant objects. Such moving parallax is another mechanism aiding depth perception in monocular vision.

Binocular vision

Vision using two eyes (binocular vision) offers several advantages over monocular vision. A larger visual field results, the blind spot and any disorder of one eye is not critical, and a vastly improved depth perception is gained. The latter is due to two factors:

(a) convergence (see page 31) whereby the visual axes automatically converge on the object studied thus permitting a judgement of distance;
(b) stereopsis (or stereoscopic vision) which, because of the separation of the eyes (about 7 cm), produces different retinal images in the two eyes. The brain then 'combines' these two different images to form a three-dimensional view of the world.

The ability of the eye to distinguish between two objects at different distances depends on the difference in the visual angle subtended by the two objects, the minimum discernible angle being about half a minute. Hence, depth perception decreases with distance, vanishing altogether beyond a certain distance (the stereoscopic range) which is normally about 60 m.

Defects of vision

Aberrations in the normal eye

Even the normal eye is subject to the inevitable aberrations present in any optical system, for example:

(a) Spherical aberration
This is the blurring of images resulting from wide beams of incident light. This clearly increases as pupil size increases and leads to greater blurring at low illuminations.

(b) Chromatic aberration
This involves the colouring and blurring of images in incident white light. Since the refractive index (and hence focal length) of the lens system depends on wavelength, different wavelengths are brought to focus at slightly different positions, leading to image blurring.

(c) Diffraction
As light passes through an aperture such as the pupil some 'spreading out' of light energy occurs. The resulting decrease in image sharpness is most pronounced with the smaller apertures used in bright light.

Long sight (hypermetropia)

Distant objects are seen clearly, giving the long-sighted eye a normal far point at infinity. However, the optical system is not powerful enough to focus near objects on to the retina (see Fig. 3.9(a)) and the near point is farther from the eye than the normal 250 mm. An additional converging lens is needed to correct the condition. For example, to bring the near point from 500 mm to 250 mm from the eye, a convex lens of focal length f mm is needed (see Fig. 3.9(b)).

$$\frac{1}{u} + \frac{1}{v} = \frac{1}{f}$$

Neglecting the distance between the corrective lens and the eye, and using the 'real-is-positive' sign convention:

$$\frac{1}{250} - \frac{1}{500} = \frac{1}{f}$$

since A is the virtual image of B in the lens.

$$\therefore \qquad f = 500 \text{ mm}.$$

Hypermetropia is common in infancy when the lens reaches adult size before the rest of the eyeball, which is consequently 'too short'.

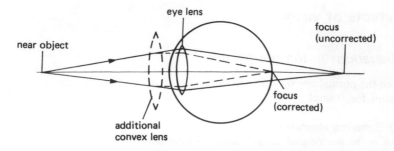

(a) Correction for long sight

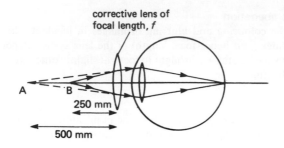

(b) Correction for near point

Figure 3.9   Long sight (hypermetropia)

Short sight (myopia)

Near objects can be seen clearly but the optical system is too powerful and focuses distant objects in front of the retina (see Fig. 3.10(a)). The far point is hence closer than infinity and a diverging lens is needed to correct the condition. To move the far point from a position P to infinity, as in Fig. 3.10(b), the focal length f mm of the diverging lens necessary is numerically equal to the distance from P to the eye. Such a lens would also move the near point farther from the eye, from D to C (Fig. 3.10(c)). If D is 200 mm from the eye, and f is -400 mm

$$\frac{1}{u}+\frac{1}{v}=\frac{1}{f}$$

$$\frac{1}{u}-\frac{1}{200}=-\frac{1}{400}$$

since D is the virtual image of C in the lens.

$\therefore$ $u=400$ mm

and the near point distance has been doubled.

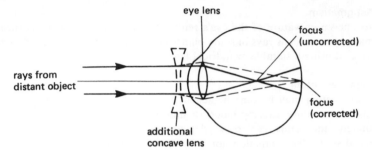

(a) Correction for short sight

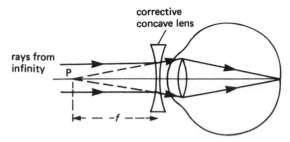

(b) Correction for far point

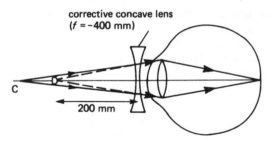

(c) Consequent change in near point

Figure 3.10   Short sight (myopia)

Other defects of vision

Other common conditions include:

(a) Presbyopia (old sight)
The lens hardens with age and becomes too stiff to accommodate. The amplitude of accommodation falls from about 14 D in young children to about 2 D at 45–50 years of age to zero above about 50 years of age. Different glasses are often required for close work and middle distance vision, and may take the form of bifocal lenses.

(b) Astigmatism

Due to uneven curvature of the eye's refracting surfaces (usually the cornea), the astigmatic eye can focus rays in some planes better than others. The condition may be corrected using a suitable cylindrical lens.

(c) Detached retina

The neural retina may become detached from the pigment epithelium (Fig. 3.2) perhaps due to injury, causing blood or fluid to collect between these two layers. Fortunately, the retina can resist degeneration for many days and may be reattached surgically to function normally again (see page 115).

(d) Colour blindness

This is due to the absence or reduced sensitivity of one or more of the three types of colour receptors. Dichromats perceive only two of the three primary colours, whilst monochromats detect shades of one colour only, since two of the three receptors are missing. The latter defect must be distinguished from total colour blindness in which the cones do not function at all and the spectral sensitivity curve is that for rods only (Fig. 3.6(a)).

Exercise 3

1 (a) Describe the optical functions of the iris and lens in the human eye. Compare them with the corresponding components in a camera.

(b) How would you expect the eye's refractive properties to change when swimming underwater? Discuss the effect, if any, of wearing goggles.

(c) The focal length, f, of a convex lens of refractive index μ_1 placed in a medium of refractive index μ_2, is given by:

$$\frac{1}{f} = k\left[\frac{\mu_1 - \mu_2}{\mu_2}\right]$$

where k is a constant. If the refractive index of the eye lens is 1.4 and its power is 18 D when in a medium of refractive index 1.33, find its approximate focal length in air.

In what ways would an air-filled eye differ from a fluid-filled one?

2 (a) What is meant by *visual acuity* and what factors tend to increase its value?

Find the minimum discernible separation of objects placed 250 mm (the normal least distance of distinct vision) from the eye, if the visual acuity is 1 per minute.

(b) A man, 70 years old, cannot see objects closer than 2 m clearly unless he uses spectacles. If he wishes to read a book held at 250 mm from his eyes, what spectacles does he need?

If his eyes can focus rays which are converging to points $\geqslant 1.5$ m behind them, calculate his range of distinct vision when wearing his spectacles.

3 (a) Explain why objects within a dark room are at first not discernible, but gradually become visible as time progresses.

(b) Why should carrots (a good source of vitamin A) supposedly help one to see in the dark?

(c) When 'bad light stops play' in a game of cricket, why should the batsman be more affected than the bowler?

(d) Why are photographic darkrooms often painted green?

4 Describe briefly how the eye perceives colour.

(a) Why does yellow light appear brighter than blue or red light of the same intensity?

(b) If an eye adapted to red light receives yellow light, describe and explain what colour it will register.

(c) When red and green lights are shone simultaneously into the eye, yellow is registered, even though there are no component yellow wavelengths present. Explain this effect.

(d) Describe and explain what happens to the perception of colour in poor light.

5 (a) With reference to the human eye, explain what is meant by *blind spot*, *accommodation*, and *near point*.

(b) Explain the following observations:
  (i) When two brightly coloured discs, one blue and one red, are projected without overlap onto a screen, most people cannot focus sharply on both at once.
  (ii) A printed page which can be read most easily when held at near point, can be read equally easily when held closer to the eye, provided it is viewed through a pinhole held close to the eye.

(c) A person with a near point 6 m from the eye needs a lens which when placed close to the eye will allow objects at a distance of 0.5 m to be seen clearly. Find the power, in dioptres, of the lens required. [JMB]

6 Give a brief account of the most common defects of vision and how they might be corrected. Diagnose the following disorders of the eye and recommend, where possible, corrective measures:

(a) far point of infinity; near point 0.75 m from the eye; amplitude of accommodation 11 D;

(b) visible spectrum width normal; difficulty in distinguishing red, orange, yellow and green;

(c) far point 2 m from the eye; near point 0.2 m from the eye;

(d) unequal focus in different planes;

(e) near point 0.5 m from the eye; amplitude of accommodation 1 D.

7 Describe the optical system of the eye and explain the meaning of far point, near point and least distance of distinct vision.

(a) A myopic child has a range of vision of 100 mm to 200 mm from the eye. What lens is necessary to enable him to see distant objects clearly? What is his new range of distinct vision when using this lens? Discuss whether his vision is likely to improve with age.

(b) A man is given two pairs of spectacles to correct his far point to infinity and his near point to 250 mm from the eye. If these spectacles have lenses with focal lengths of -2 m and 0.5 m respectively, find the far and near points of his unaided eyes.

8 Discuss the mechanisms responsible for the perception of depth in a scene and indicate which ones are not applicable when using only one eye.

A person can focus objects only when they lie between 0.5 m and 3 m from his eyes. Find his range of distinct vision when he is wearing spectacles:

(a) to correct his far point to infinity;

(b) to correct his near point to 250 mm from the eye.

9 Figure 3.11 represents the appearance of the pupil and iris in conditions of different light intensities.

(a) Label the pupil and the circular and radial muscles of the iris.

(b) Identify which illustration refers to bright light conditions, normal illumination, and dim light.

(c) Explain how this brightness control mechanism works.

(d) Discuss briefly any further techniques employed by the eye to cope with varying incident light intensities.

(i)　　　　　　　　　　(ii)　　　　　　　　　　(iii)

Figure 3.11.

10 Either:

(a) Describe the trichromatic theory of colour vision. How does the spectral sensitivity of the normal eye vary with intensity and wavelength of incident light?

or:

(b) Discuss the various reflex mechanisms operating in the eye and indicate any connections there might be between them.

A piece of transparent paper with an opaque cross on it is held about 20 cm in front of the eye of a subject, who then gazes through the paper at some distant object. When the subject suddenly looks at the cross, describe and explain the changes which are *seen* to occur in his eye.

11 (a) (i) State the nature of the defect known as astigmatism and name the type of lens used to correct it.

(ii) An old person and a young person each wear spectacles with converging lenses. In *each* case name the probable eye defect, state its nature and how correction is achieved.

(b) The combined power of cornea and lens of a normal, unaccommodated eye is 59 dioptres. If the eye focuses on an object 250 mm away find the change in power, in dioptres, of the eye.

A myopic eye has a far point of 1.0 m and a near point of 150 mm. State the type of spectacle lens needed for viewing an object at the normal far point and determine its power, in dioptres.

Find the near point when wearing this lens.　　　　　　　　[JMB]

4 | The physics of hearing

Properties and propagation of sound waves

Sound waves in a medium

A sound wave is a longitudinal pressure wave propagated by oscillations of the particles of the medium through which it travels. The velocity c of such a wave through a medium of density ρ and modulus of elasticity E is given by:

$$c = \sqrt{\frac{E}{\rho}}$$

The medium offers an opposition to the passage of sound waves through it, analogous to the opposition to the flow of current in an electrical circuit. Just as electrical impedance is used in the latter case, acoustic impedance is used to describe the opposition of an elastic medium to sound waves. The specific acoustic impedance Z of a medium is found to be given by:

$$Z = \rho c \qquad [4.1]$$

The intensity I of a sound wave in a medium may be shown to be:

$$I = \frac{1}{2}\frac{a^2}{\rho c} = \frac{a^2}{2Z}$$

where a is the amplitude of the sound wave. Hence, for a sound wave in a given medium:

$$I \propto a^2$$

The progressive loss in intensity of a sound beam as it traverses a medium is known as attenuation. It results from absorption and deviation from the beam direction through processes such as divergence, scattering, diffraction, and so on. Like the attenuation of other wave forms (e.g. X-rays) the attenuation of sound waves is exponential, that is:

$$I_x = I_0 e^{-\mu x}$$

where μ is a constant, known as the attenuation coefficient. Absorption alone also results in an exponential loss in intensity, characterised by an absorption coefficient, k:

$$I_x = I_0 e^{-kx} \qquad [4.2]$$

Reflection and transmission at boundaries

The regular reflection and refraction of sound waves at a boundary between media obey the same laws as light waves (see Fig. 4.1). Even at normal incidence a certain amount of energy is reflected back, the amount depending on the acoustic

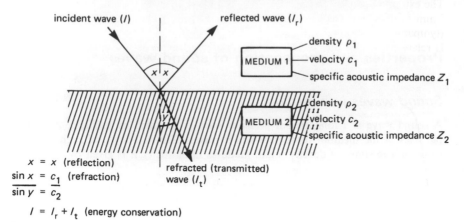

$x = x$ (reflection)

$\dfrac{\sin x}{\sin y} = \dfrac{c_1}{c_2}$ (refraction)

$I = I_r + I_t$ (energy conservation)

Figure 4.1   Reflection and refraction of sound waves

impedances of the two media. The fraction of the energy reflected back (I_r) to that transmitted (I_t) is known as the intensity reflection coefficient α_r and is given by:

$$\alpha_r = \frac{I_r}{I_t} = \left(\frac{Z_2 - Z_1}{Z_2 + Z_1}\right)^2 \qquad [4.3]$$

(a) When $Z_1 = Z_2$, $\alpha_r = 0$. There is no reflection, the wave is totally transmitted, and there is said to be a good acoustic match between the two media.
(b) When $Z_1 \ll Z_2$ or $Z_2 \ll Z_1$, $\alpha_r \to 1$. Most of the incident energy is reflected, and there is said to be an acoustic mismatch between the media. This is typically the case when dealing with interfaces between a gas and a liquid or solid, since the density (and hence Z) of the gas is so much smaller than that of either a liquid or a solid.

Characteristics of a sound

The perception of sound depends on certain properties which 'characterise' a note:

(a) pitch, determined by frequency;
(b) loudness, determined by intensity and frequency;
(c) quality (or timbre), determined by the frequencies present and their relative amplitudes.

In general, musical notes are regular mixtures of a fundamental frequency and its harmonics. Noise, on the other hand is a random mixture of unrelated frequencies.

The structure of the ear

The external or outer ear

The external auditory passage (Fig. 4.2) is an S-shaped tube about 3 cm long and 7 mm in diameter which links the pinna at the side of the head to the ear drum (tympanic membrane). Although the ear drum is only about 0.1 mm thick, it is relatively strong due to the combination of radial and concentric fibres.

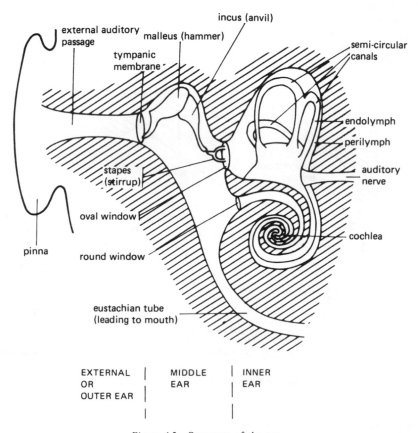

Figure 4.2   Structure of the ear

The middle ear

This consists of an air-filled cavity (the tympanic cavity) in which is suspended by ligaments a chain of three small bones, called the malleus, incus and stapes. These link the ear drum to the oval window, a membrane-covered opening into the inner ear.

The inner ear

The complex bony chamber of the inner ear is filled with perilymph, a fluid similar in composition to cerebrospinal fluid, having a low potassium ion concentration and a high sodium ion concentration. Within the inner ear chamber is a membranous labyrinth, forming the semi-circular canals and cochlear duct which contains a different fluid, the endolymph. This is more like the fluid in cells having a high potassium ion concentration and a low sodium ion concentration.

The semi-circular canals are concerned with the detection of movement of the body and balance, and do not contribute to the process of hearing. The cochlea, on the other hand, is the most delicate and intricate organ in the acoustic chain. It is a spirally coiled tube, of total diameter about 3 mm and volume about 100 mm^3 (approximately the volume of two drops of water), and has a blind apex. Running along its entire length are two membranes, the basilar and vestibular membranes, which divide the cochlea into three cavities (Fig. 4.3), the scala vestibuli, the scala

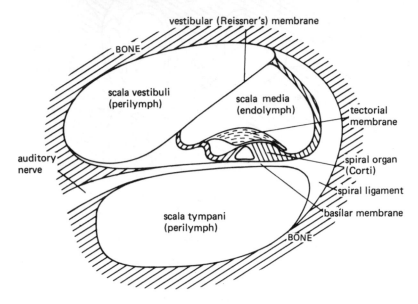

Figure 4.3   Cross-section of the cochlea

tympani and the scala media. The scala media is the membranous cochlea or cochlear duct, containing endolymph, and it terminates just before the apex of the cochlea, thus providing communication between the scalae vestibuli and tympani through the small opening known as the helicotrema (Fig. 4.4). The scala tympani communicates with the tympanic cavity via the membrane-covered round window, and the scala vestibuli opens through the oval window.

The basilar membrane is composed of about 20 000 or more largely non-cellular fibres, ranging from short, tense fibres about 0.04 mm long at its base near the oval window, to long, slack fibres about 0.5 mm long near the helicotrema. Located on top of the basilar membrane, and also running the entire length of the cochlear

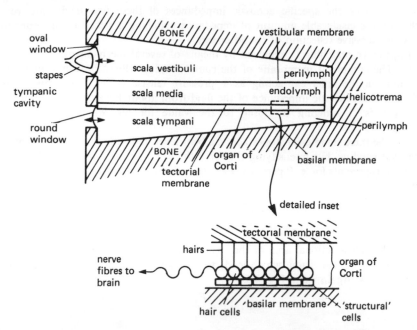

Figure 4.4   Diagrammatic longitudinal section of the cochlea (uncoiled)

duct, is a complex structure called the organ of Corti. This consists of supporting or 'structural' cells forming a relatively rigid framework above the basilar membrane, and sensory or hair cells communicating with the brain via nerve fibres.

The ends of delicate hairs which protrude from the hair cells are embedded in a rather rigid, massive structure called the tectorial membrane. The hair cells have a negative intracellular potential of about −70 mV with respect to the perilymph. Furthermore, the resting endolymph has a potential of about +80 mV with respect to the perilymph, this being known as the endocochlear potential. Thus, the upper borders of the hair cells, which project into the endolymph of the scala media, have a total membrane potential (potential difference across the cell membrane) of about 150 mV.

The functioning of the ear

Outer and middle ear

The external ear collects sound waves and transmits them (still as air vibrations) to the ear drum, which is then forced to vibrate in step. This in turn causes the associated malleus to vibrate accordingly. These mechanical vibrations, or thrusts, are then transmitted by the incus and stapes to the membrane across the oval window, and thence to the fluid of the inner ear.

Vibrations are thus transmitted from air in the outer ear to a fluid in the inner ear, and since the specific acoustic impedances of these two media are very different, a considerable amount of impedance matching is necessary in order to prevent excessive reflection of sound energy.

Impedance matching is achieved by means of several mechanisms.

(a) The compliant membrane of the round window allows bulk movement of the fluid in the inner ear acting as a 'pressure release valve'. This reduces the effective characteristic impedance of the fluid from about $10^6 \, \text{kgm}^{-2}\text{s}^{-1}$ to about $10^5 \, \text{kgm}^{-2}\text{s}^{-1}$, bringing it closer to the value for air, which is just over $400 \, \text{kgm}^{-2}\text{s}^{-1}$

(b) The three middle-ear bones act as a bent lever, reducing the displacement at the oval window, but intensifying the force there. Using the notation of Fig. 4.5, where F represents force, P pressure, l length and A area, taking moments about O gives:

$$F_1 l = F_2 . 2l/3$$

$\therefore$
$$F_2 = 3F_1/2 \text{ (i.e. force amplification)}$$

(c) The cross-sectional area of the oval window is about twenty times smaller than that of the ear drum, which leads to a magnification of pressure across the oval window. Again using Fig. 4.5:

$$P_1 - P = \frac{F_1}{A_1}$$

and
$$P - P_2 = \frac{F_2}{A_2}$$

$\therefore$
$$\frac{P - P_2}{P_1 - P} = \frac{F_2}{F_1} . \frac{A_1}{A_2}$$

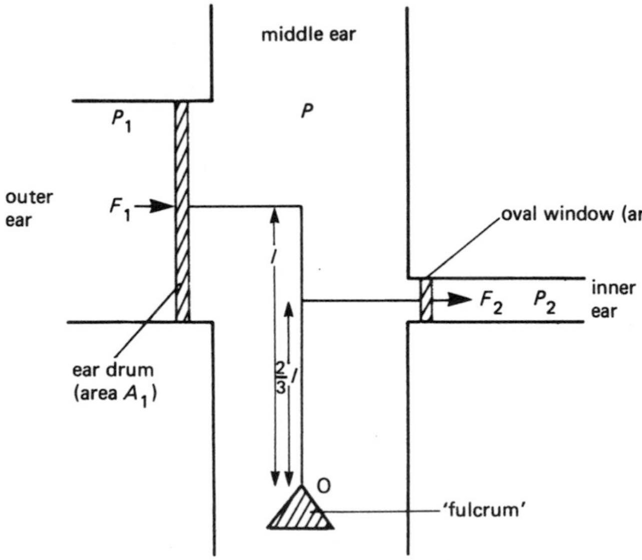

Figure 4.5   The middle ear lever system

Since $A_1 \approx 0.6\,\text{cm}^2$ and $A_2 \approx 0.03\,\text{cm}^2$:

$$\frac{P-P_2}{P_1-P} \approx \frac{3}{2} \times 20 = 30 \qquad \text{(i.e. pressure amplification)}$$

Thus the higher pressures necessary to put into motion the denser fluid of the inner ear are provided, and the effective mismatch in acoustic impedances between the outer and inner ears is reduced.

The outer and middle ear also serve to protect the delicate inner ear as well as the ear drum. In particular, the eustachian tube linking the middle ear cavity to the mouth (and hence the atmosphere) permits the equalising of pressure on both sides of the ear drum and so prevents it rupturing in cases of large pressure differences.

Inner ear

Vibrations of the membrane across the oval window produce corresponding pressure waves in the perilymph in the scala vestibuli (Fig. 4.4). At frequencies below about 20 Hz the pressure waves cause the perilymph to move back and forth through the helicotrema, producing little effect on the basilar membrane. As the frequency rises above about 20 Hz, however, the inertia of the fluid dictates that the pressure waves 'short circuit' through the basilar membrane, causing the latter to vibrate.

The exact process whereby basilar membrane vibrations are translated into neural impulses is not yet fully understood. However, it is widely accepted that movement of the basilar membrane causes the hairs to bend back and forth, and this distortion stimulates the hair cells to initiate neural impulses which travel along the auditory nerve to the brain. For example, the distortion may modify the membrane permeability of the hair cells resulting in ionic composition and membrane potential changes, and the consequent initiation of neural impulses. (The conversion of mechanical distortions to electrical signals is also observed in piezoelectric crystals, see Chapter 10).

The ear's frequency response

Resonance in the outer and middle ear

Resonance is the selective reinforcement of certain frequencies, determined by the particular structure of the vibrating system. The external auditory passage acts as an air column resonator, and exhibits a slight resonance at about 3000 Hz. The middle ear displays a broad, but again slight, resonance between about 700–1400 Hz, the effect being at its greatest at about 1200 Hz. In general, sound transmission to the cochlea is excellent between about 600–6000 Hz, but deteriorates outside these limits.

Frequency discrimination in the inner ear

As the foot of the stapes moves inward, initiating a pressure wave in the perilymph, the basilar membrane near the oval window is forced to bulge towards the round window, which in turn bulges outward. The elastic tension thus built up in the basilar fibres near the oval window causes a travelling wave to propagate along the membrane (Fig. 4.6.).

This travelling wave has a wavelength and velocity which decrease with distance from the oval window but, perhaps even more important, its amplitude also varies reaching a maximum at some well-defined point on the membrane before rapidly decaying to zero. The exact location of the maximum is determined by the stimulating frequency: low frequencies produce peaks far from the oval window, where the basilar fibres are longer and the total mass of perilymph involved in vibrations is greatest; high frequencies give rise to maxima near the oval window, where basilar fibres are shorter, narrower and more rigid, and the amount of perilymph set in vibration small. Thus each point along the basilar membrane becomes associated with a particular frequency, as does its corresponding neuron. The brain can then discriminate frequencies since specific neurons are activited by particular frequencies.

Although this 'place principle' is widely accepted as the primary method of frequency discrimination, there is some indication that a 'volley' mechanism predominates at frequencies below about 120 Hz. According to this theory, movement of the basilar membrane initiates volleys of. neural impulses, the frequency of firing being synchronous with the frequency of the incoming sound waves.

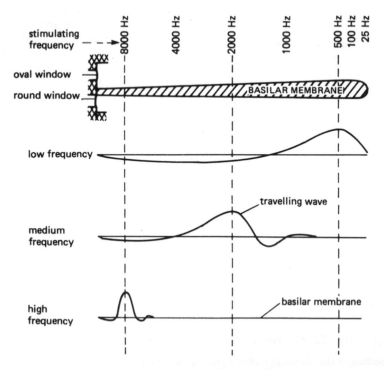

Figure 4.6   Frequency discrimination

Frequency ranges

The range of audible frequencies varies considerably from person to person, but the average range is 20–20 000 Hz.

Below 20 Hz, sound waves are termed subsonic or infrasonic, and merely cause the perilymph in the cochlea to move back and forth through the helicotrema without disturbing (and hence stimulating) the basilar membrane and associated nerve fibres.

Basic speech frequencies range from about 60–500 Hz (omitting harmonics); a good soprano can reach about 1300 Hz; and the piano ranges from about 25–4000 Hz. Since the quality of a sound depends on the presence of harmonics, however, it is necessary for the ear to detect higher frequencies. The upper audible frequency limit depends very much on the power of the source and the age of the observer. Few people can detect low-power sounds of frequencies greater than about 12 000 Hz. On the other hand, high-power sounds of greater than 21 000 Hz can be detected by some people, particularly young children. Fig. 4.7. shows how audible frequency ranges depend on the incident sound intensity.

Sound waves of frequencies greater than 20 000 Hz are called ultrasonic (see Chapter 10).

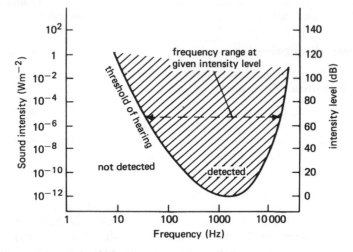

Figure 4.7   The average audible range

Frequency differences

In the range 60–1000 Hz, frequencies differing by 2–3 Hz can be distinguished when presented separately. Beyond 1000 Hz, the ability to discriminate close frequencies decreases by about 2 Hz per 1000 Hz. Above about 10 000 Hz all frequencies are poorly discriminated.

This non-linearity in the ear's frequency response is reflected in the development of musical intervals, which depend on the ratios of frequencies rather than their differences. The musical interval between two notes is an upper octave if the ratio of their frequencies is 2:1. For example, a note of frequency 1024 Hz is an octave above a note of frequency 512 Hz, which in turn is an octave above a note of frequency 256 Hz (middle C).

If two notes of similar loudness and frequency are heard simultaneously, a periodic rise and fall in intensity is heard, a phenomenon known as beats. The beat frequency is the number of intense sounds heard per second and is found to be equal to the difference between the two component frequencies. The ear can detect beats of frequency up to about 6 Hz.

Sensitivity of the ear

Threshold of hearing

The smallest sound intensity discernible by the ear is known as the threshold of hearing, or threshold of audibility. It depends on the frequency of the incoming signal (see Fig. 4.7) reaching its lowest value of about 10^{-12} W m^{-2} at a frequency around 2–3 kHz. If the threshold was much below this value, the random thermal motion of air molecules in the atmosphere would be (undesirably!) detected. Furthermore, if the ear was as sensitive at low frequencies as it is around 2 kHz, many physiological noises, such as blood flow, would be discernible.

Loudness discrimination

The brain perceives an increase in intensity of a sound of given frequency as an increase in loudness. This is due to at least three mechanisms:

(a) the amplitude of vibration of the basilar membrane and hair cells increases producing a greater stimulation of nerve endings;
(b) more nerve fibres become stimulated at the fringes of the vibrating portion of the membrane, due to its increased amplitude;
(c) certain hair cells, having higher thresholds for activation, become stimulated.

Thus more nerve fibres are stimulated to a greater degree, increasing the nerve impulse frequency to the brain which then registers a louder sound.

Although loudness is closely connected with the intensity of the incident sound wave, the two quantities are not equivalent. Intensity is a physically defined quantity, independent of the observer, whereas loudness is a subjective sensation, depending on the exact transfer characteristics from ear drum to neural impulses for each individual ear.

Loudness, like the threshold of hearing, is strongly dependent on frequency. Fig. 4.8 shows some curves of equal loudness, measured in phons (see page 60) and illustrates that sounds of equal loudness do not necessarily correspond to equal intensity levels.

The ear's logarithmic response

The ear can detect a tremendous range of intensities, from a threshold of about 10^{-12} W m^{-2} to an upper limit of about 100 W m^{-2}, at which discomfort is experienced, and beyond which permanent damage results.

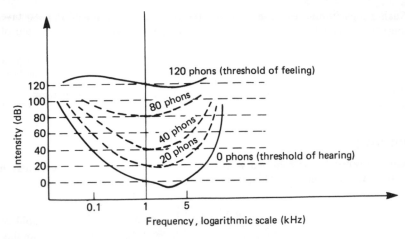

Figure 4.8 Curves of equal loudness

The detection is not, however, linear. Small changes in intensity are more easily discerned at lower intensities than at higher intensities, so that equal changes in intensity across the range are not registered as equal changes in loudness. For example, if the intensity of sound from a source of given frequency increases from $1 \times 10^{-5} \, \text{W m}^{-2}$ to $2 \times 10^{-5} \, \text{W m}^{-2}$, and then from $2 \times 10^{-5} \, \text{W m}^{-2}$ to $4 \times 10^{-5} \, \text{W m}^{-2}$, the loudness of sound appears to the ear to increase in equal steps. It is found that

$$\text{loudness change} \propto \frac{\text{intensity change}}{\text{initial intensity}}$$

$$\therefore \qquad \mathrm{d}L \propto \frac{\mathrm{d}I}{I}$$

$$\therefore \qquad \mathrm{d}L = k\frac{\mathrm{d}I}{I}$$

where L represents loudness, I represents intensity and k is a constant. Integrating gives:

$$L = k \log_e I + C$$

where C is a constant determined by the following boundary conditions: when $I = I_0$ the threshold intensity, $L = 0$, giving,

$$C = -k \log_e I_0$$

$$\therefore \qquad L = k(\log_e I - \log_e I_0)$$

$$\therefore \qquad L = k \log_e \left(\frac{I}{I_0}\right)$$

$$\therefore \qquad L \propto \log_{10}\left(\frac{I}{I_0}\right) \qquad\qquad [4.4]$$

The ear thus displays a logarithmic response, the relationship [4.4] being an example of the Weber–Fechner law.

Such a logarithmic response is always the result of a relationship of the form: change in perceived property $\propto$ fractional change in stimulus

$$\text{eg. } dL \propto \frac{dI}{I}$$

and is displayed by many other sensory experiences, including the perceptions of brightness and weight.

Sensitivity

Maximum sensitivity is achieved when the smallest discernible relative change in intensity, $\Delta I/I$ is minimum. Hence, sensitivity S is defined by:

$$S = \log\left(\frac{I}{\Delta I}\right)$$

S depends strongly on frequency, being maximum at about 2000 Hz, (see Fig. 4.9).

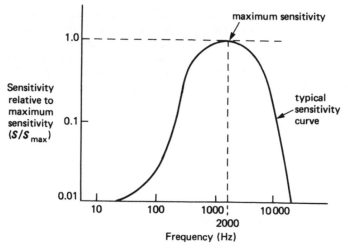

Figure 4.9   Sensitivity of the ear

Units of sound

Intensity level and the bel

The intensity level (or relative intensity) of a signal is its intensity relative to an agreed 'zero' of intensity. The latter is chosen as the signal intensity I_0 at the threshold of hearing which is generally accepted as $10^{-12}\,\text{W m}^{-2}$ ($1\,\text{pW m}^{-2}$) at 1 kHz.

Since the ear displays a logarithmic response to intensity, (see equation [4.4]), it is convenient to define the unit of intensity level using a logarithmic scale. The intensity level of a signal of intensity $I\,\text{W m}^{-2}$, measured in bels (B) is then defined by:

$$\text{intensity level} = \log_{10}\left(\frac{I}{I_0}\right) \text{B}$$

For example, a sound of intensity $10^{-11}\,\text{W m}^{-2}$ has an intensity level of: $\log_{10}(10^{-11}/10^{-12})\,\text{B}$ or 1B.

The decibel

The bel is a large unit, representing intensity differences in the ratio 10:1. More commonly used is the decibel (dB) where:

$$1\,\text{B} = 10\,\text{dB}$$

and

$$\text{intensity level} = 10\,\log_{10}\left(\frac{I}{I_0}\right) \text{dB}$$

For example, relative to a threshold of $10^{-12}\,\text{W m}^{-2}$, a sound of intensity $10^{-5}\,\text{W m}^{-2}$ has an intensity level of:

$$10\,\log_{10}\left(\frac{10^{-5}}{10^{-12}}\right) = 70\,\text{dB}$$

The difference in intensity level between two sounds of intensities I_1 and I_2 is given by:

$$10\left[\log_{10}\left(\frac{I_2}{I_0}\right) - \log_{10}\left(\frac{I_1}{I_0}\right)\right]\text{dB}$$

$$= 10\,\log_{10}\left(\frac{I_2}{I_1}\right)\text{dB}$$

For example, the difference in intensity level between a sound of intensity $10^{-6}\,\text{W m}^{-2}$ due to normal speech and one of intensity $10^{-1}\,\text{W m}^{-2}$ in a submarine engine room is:

$$10\,\log_{10}\left(\frac{10^{-1}}{10^{-6}}\right)\text{dB}$$

$$= 50\,\text{dB}$$

The minimum change of intensity level which the ear can detect is on average about 1 dB although the value can range from about 0.3–5 dB depending on frequency and absolute sound level; 1 dB corresponds to a ratio of sound intensities I_2/I_1 given by:

$$1 = 10\,\log_{10}\left(\frac{I_2}{I_1}\right)$$

$$\therefore \quad \frac{I_2}{I_1} = \text{antilog}_{10}\left(\tfrac{1}{10}\right)$$

$$\therefore \quad \frac{I_2}{I_1} = 1.26$$

This is equivalent to a difference in intensity of about 25 per cent. (This minimum discernible relative change does not compare favourably with values for other senses, e.g. feeling of weight $\approx$ 10 per cent, estimating length of lines $\approx$ 2 per cent, and brightness of light $\approx$ 1 per cent).

The phon

It might appear that loudness ($L \propto \log_{10} I/I_0$) and intensity level ($\log_{10} I/I_0$) are equivalent quantities, both of which might be measured in B or dB. This, however, is not so, since intensity level is a physically-defined quantity, independent of the observer, whereas loudness is a subjective sensation and strongly dependent on frequency (Fig. 4.8). Thus, in order to obtain a consistent unit for loudness, a standard frequency of 1 kHz is chosen and all sounds are compared at this frequency. To do this, the source of unknown loudness is placed near the standard source of frequency 1 kHz, and the power of the standard source is adjusted until the two sources sound equally loud. If the intensity level of the standard source is then n dB, the loudness of the unknown source is said to be n phons.

Noise and its physiological effects

Noise is a complex disturbance, having many component frequencies and producing varying effects on different individuals. Table 4.1 illustrates some common sources of noise and their average intensity levels, sometimes referred to as noise levels.

Table 4.1 Common sources of noise

Source	Intensity level (dB)	Sound intensity (W m^{-2})
'Silence' (Threshold of hearing)	0	10^{-12}
Library	20	10^{-10}
Average home	30	10^{-9}
Background music	40	10^{-8}
Speech at 0.6 m	60–80	10^{-6}–10^{-4}
Heavy traffic	80	10^{-4}
Pneumatic drill	90	10^{-3}
Factory	80–130	10^{-4}–10
Jet overhead	100	10^{-2}
Thunder overhead	110	10^{-1}
Threshold of feeling	120	1

High intensity noise is generally accepted as greater than 85 dB, and as the level increases there are a number of harmful results:

(a) change in hearing acuity and possible damage to the cochlea;
(b) stimulation of receptors in the skin;
(c) significant changes in pulse rate;
(d) vibrations of muscles and incoordination;

(e) feelings of fear, annoyance, dissatisfaction;
(f) inability to perform skilled and unskilled tasks;
(g) nausea, vomiting, dizziness ($>130\,$dB);
(h) pain in middle ear ($\approx 140\,$dB);
(i) temporary blindness ($>140\,$dB);
(j) mild warming of body surfaces ($>150\,$dB);
(k) minor permanent damage if prolonged ($\approx 160\,$dB);
(l) major permanent damage in a short time ($\approx 190\,$dB).

Stereophonic hearing

The process by which a source of sound is localised without visual aid is complex and involves several mechanisms:

(a) The time lag between the arrival of the sound at the two ears is an accurate method, important particularly for brief click-like sounds and for frequencies below about 3 kHz. However, there is some frequency dependence and confusion can arise between locations giving rise to the same time lag. A quick rotation of the head, however, can often remove such uncertainty.
(b) The phase relationship between the signals entering the two ears is significant for low-frequency, smoothly changing types of sound, and together with (a) is very important in localising the human voice.
(c) The difference in intensity of the sound entering the two ears is important. The head acts as a good absorber and tends to 'shadow' sounds. The effect is more prominent at high frequencies, when diffraction effects are low and a smaller fraction of the incident sound energy consequently reaches the 'shielded' ear. Intensity level differences of about 30 dB can then be experienced, and the method becomes particularly accurate above about 3 kHz.

In general, high-frequency sounds are more easily located than low-frequency sounds, largely due to increased diffraction at low frequencies causing a more diffuse sound.

Exercise 4

1 Explain the meaning of the terms: (a) *acoustic impedance*, and (b) *intensity reflection coefficient*.

A source of sound is situated between an observer and a flat wall. If the intensity reflection coefficient of the wall is 0.8, find the difference, in dB, of the intensity levels of the sound heard by the observer directly and by reflection. State any assumptions you make.

2 What is meant by *acoustic mismatch* and of what relevance is it to the functioning of the ear?

If the specific acoustic impedance of air is $420\,$kg m^{-2} s^{-1} and that of water is $1.5 \times 10^6\,$kg m^{-2} s^{-1}, calculate the fraction of sound intensity reflected back to that transmitted at an air/water interface.

The impedance matching between sound waves in air in the outer ear and fluid vibrations in the inner ear is about 50–75 per cent perfect for frequencies between about 300–3000 Hz. How can you explain this?

3. What are the three characteristics of a note? Describe carefully how one of them is discriminated by the ear.

The intensity of the noise from an engine of a jet plane is $0.4 \, W \, m^{-2}$ at a distance of 10 m. When the plane is directly overhead at an altitude of 200 m, estimate:
(a) The intensity of the noise heard;
(b) The noise level in dB, relative to a threshold of $10^{-12} \, W \, m^{-2}$;
(c) The reduction in noise level from that experienced at 10 m from the jet.

4 Would you describe the human perception of
(a) pitch, (b) loudness
    as linear responses? Give reasons.
(c) If an amplifier increases the intensity of sound from $4 \, \mu W \, m^{-2}$ to $5 \, \mu W \, m^{-2}$, express the amplification in dB.
(d) Noise coming through an open window at 70 dB is reduced to 50 dB when the window is closed. What percentage of the sound is excluded?

5 Distinguish between the intensity, intensity level and loudness of a sound and define the units used to measure them. Express the following in decibels:
(a) the increase in intensity level due to the power from a loudspeaker changing from 5W to 20W;
(b) the decrease in intensity level due to moving from 5 m to 15 m away from a source, which is emitting sound energy uniformly in all directions;
(c) the difference in intensity level between a noise of $10^{-3} \, W \, m^{-2}$ from a 'tube' train and a noise of $5 \times 10^{-6} \, W \, m^{-2}$ due to shouting at a distance of about 4 m.

6. What is meant by stereophonic hearing? Describe the mechanisms by which it is achieved and discuss whether these are more effective at high or low frequencies.

A loudspeaker produces a sound intensity level of 6 dB above a certain reference level at a point P, 20 m from it. Find:
(a) The intensity level at a point 30 m from the loudspeaker;
(b) The intensity level at P if the electrical power to the loudspeaker is halved.

7 (a) Explain what is meant by the decibel scale for comparing two quantities, and give a definition of a reference level for such a scale for sound intensities.
(b) A listener wears headphones connected to the output of a stereo amplifier whose output is initially 2 mW. The listener slowly increases the output power and subjectively does not discern an increase in sound intensity until the power has risen to 2.5 mW. Successively discernible increases then occur at 3.2 mW and 4.0 mW. Use these results:
    (i) to show why the decibel scale is a useful one;
    (ii) to calculate the ratio of the amplitudes of the pressure waves of two sounds whose difference in intensity is just discernible.   [JMB]

8 Describe the sensitivity of the ear. Discuss any special features which optimise the ear's response.

How would you use the inverse square law to 'protect' yourself from an intense source of noise? A man is standing 3 m from a siren when it goes off. How far will he have to move away from the siren to reduce the intensity level by 20 decibels? (You may assume the siren emits sound energy uniformly in all directions.)

9 (a) Explain what is meant by the *intensity of sound* and indicate how the loudness perceived is related to the intensity received at the ear.
    Calculate the intensity of the loudest sound the ear can withstand given that the intensity level, referred to a threshold for human hearing of $10^{-12}\,\mathrm{W\,m^{-2}}$ of the same sound is 120 dB.

   (b) The eye has a threshold for perception when 100 photons per second, of wavelength 510 nm, enter the pupil. The effective area of the external entrance of the auditory canal (auditory meatus) is 65 mm^2.
    If the threshold sensitivity of either organ is defined as the least power required to produce a perceptible signal, calculate the ratio of the threshold sensitivity of the eye to that of the ear.
    Comment on the significance you think this result has for man.
    Planck's constant $= 6.6 \times 10^{-34}\,\mathrm{Js}$
    Speed of light $= 3.0 \times 10^8\,\mathrm{m\,s^{-1}}$ [JMB]

10 (a) With the aid of a labelled diagram of the middle ear, explain how sound energy is transmitted across the tympanic cavity. Give a reason why the pressure changes due to sound are increased as a result of this transmission and state an approximate value for the increase.

   (b) A meter, which measures relative intensity level of sound referred to $1\,\mathrm{pW\,m^{-2}}$, records a value of 97 dB when a pneumatic drill is switched on some distance away. Calculate the intensity of sound at the meter.
    A second drill, identical to the first, is placed close to it. State the intensity of sound at the meter when both drills are working and hence calculate the *increase* in the meter reading.
    Explain why it is convenient to use the decibel scale for such measurements.
                                                                                       [JMB]

11 (a) Compare what is seen by an observer, with normal eyesight, when a coloured object is illuminated by white light of (i) high intensity and (ii) low intensity. Give an explanation for your answer in terms of the behaviour of the eye.
    Sketch a graph showing the spectral response of the eye.

   (b) Two neighbouring, independent point sources of light are just resolved visually under optimum viewing conditions when they subtend 0.3 milliradians at the eye. If the sources are 0.5 m from the eye calculate their separation. Explain, in terms of the structure of the retina, how the two retinal images must be positioned to be seen as separate.

   (c) Explain what is meant by *intensity of sound*. Sketch a graph showing the frequency response of the ear in a healthy young person, indicating approximate numerical values on the axes.
    Describe how the graph might be expected to differ for an older person.
                                                                                       [JMB]

PART II
Biomedical measurement

5 | Electrical conductance

The origin of biopotentials and their detection

Biopotentials

A biopotential is a potential generated inside the body and generally arises from salt concentration differences across cell membranes. These so-called 'membrane potentials' are exhibited by nerve, muscle and gland cells.

Although all living cell membranes pass water, the solute transmitted depends on the state and type of membrane. For example, cell membranes of the large intestine pass even large molecules, whereas nerve fibre cell membranes only pass the ions of NaCl and KCl (i.e. Na^+, Cl^- and K^+).

Such a nerve fibre, or axon, is illustrated in Fig. 5.1. It is a long, thin (a few

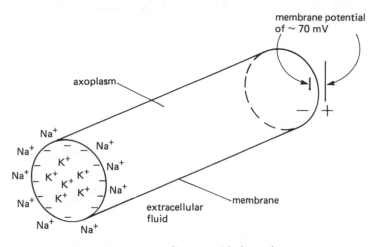

Figure 5.1   A nerve fibre (axon) in its resting state

micrometres in diameter) extension of a nerve cell (neuron), and consists of a central core of axoplasm, surrounded by a high-resistance membrane. In its resting state, there is a high concentration of negative ions and K^+ ions inside, and Na^+ ions outside the membrane, the balance being maintained by osmotic and mechanical forces. A resting membrane potential of about 70 mV is hence established as shown.

When a nerve, muscle or gland cell responds to a stimulus, the membrane potential exhibits a series of reversible changes, namely depolarisation, reverse polarisation and finally repolarisation. This sequence constitutes an 'action potential'.

Action potentials

When a nerve cell, for example, is stimulated, the cell membrane suddenly becomes permeable to Na^+ ions which then move into the axoplasm from their higher concentration area outside (see Fig. 5.2). The increase in positive charge inside the

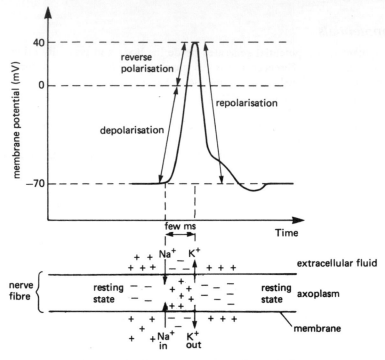

Figure 5.2   The action potential of a nerve fibre

cell leads to a change in the membrane potential from about $-70\,mV$ to $0\,mV$ (depolarisation) and further to about $+40\,mV$ (reverse polarisation). Almost immediately, the membrane becomes impermeable to Na^+ ions and permeable to K^+ ions, which consequently leave their high concentration area inside the fibre and move out, thereby restoring the original membrane potential of $-70\,mV$, (repolarisation). The Na^+ and K^+ ions are re-exchanged later during a slower recovery period.

The action potential of heart muscle, on the other hand, is shown in Fig. 5.3. Each complete cycle of depolarisation (causing contraction), reverse polarisation and repolarisation (relaxation) corresponds to one heartbeat.

The depolarised (active) region of a fibre acts as a trigger, stimulating the

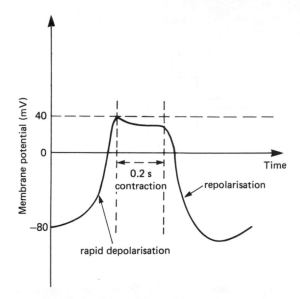

Figure 5.3   The action potential of heart muscle

adjacent region to follow through the same action potential. The rate of propagation of such an action potential depends on several factors including type of cell, fibre diameter and temperature. For example, propagation along nerve fibres may reach 150 m s^{-1}, whereas the transmission along muscle fibres is much slower.

Detection of biopotentials

Electrodes placed on the surface of the body measure the coordinated activity of a large group of cells and thus register the local action potential, which may be neural or muscular, depending on the electrode locations.

Since body materials, particularly skin, are poor electrical conductors, (resistivities range from about $0.6 \, \Omega\text{m}$ for plasma to about $25 \, \Omega\text{m}$ for fat), the selection and preparation of electrode site are most important. For instance, a suitable site must be well-cleaned, hair-free and rubbed to remove some outer cells. A conductive, yet non-irritant, electrode paste is rubbed into the site to improve the electrical contact and the applied electrode is then held firmly in place using tape.

Electrocardiography

Action of the heart

The heart (Fig. 5.4) consists of four chambers, namely the right and left auricles or atria which receive incoming blood, and the right and left ventricles which expel

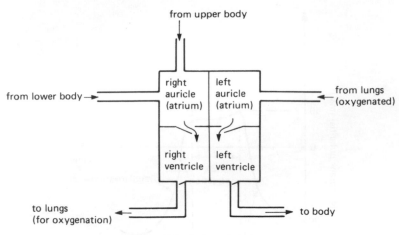

Figure 5.4   Action of the heart

blood from the heart. It beats about 60–80 times per minute, each beat being electrically triggered by a pulse arising in the upper right region of the heart. This pulse spreads across the atria causing them to contract and so force blood into the ventricles. Shortly after, the pulse reaches the ventricles causing them to contract and so pump blood out of the heart. The presence of one-way valves ensures that blood is pumped in the right direction.

The electrocardiogram (ECG)

During each heartbeat the spread of action potentials results in the formation of biopotentials between depolarised cells and those awaiting depolarisation. Using metal electrodes on the body surface, such biopotentials, though somewhat reduced, may be detected and subsequently amplified and recorded. Such a recording is known as an electrocardiogram (ECG) and its shape depends on the condition of the subject and the location of the electrodes.

A typical ECG is shown in Fig. 5.5 and shows the following features:

(a) The P–wave occurs during depolarisation of the atria which causes atrial contraction;

(b) the QRS pulse corresponds to depolarisation (and subsequent contraction) of the ventricles;

(c) The T wave occurs during ventricular repolarisation which corresponds to the relaxation of the ventricles.

The repolarisation of the atria does not normally appear in the ECG since it coincides with, and is obscured by, the large QRS pulse.

The ECG may be recorded using a chart recorder, or displayed after amplification on a cathode ray oscilloscope (CRO).

Since all muscular activity gives rise to small biopotentials any other muscular movement must be minimised, to prevent the appearance of spurious signals or 'artifacts'. Patient relaxation is very important in this respect.

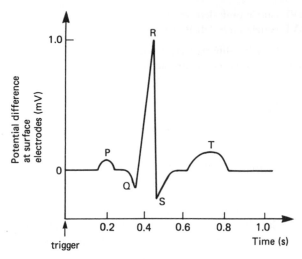

Figure 5.5   A typical ECG

ECG electrode sites

The ECG electrodes may be placed over the heart, or on the torso and limbs at points where the major arteries run close to the surface. The two major types of ECG connections are termed bipolar leads and unipolar leads.

(a) Bipolar leads

These record the potential difference between two points on the body surface, typical sites being the two arms and the left leg. A differential amplifier is employed to register the difference in voltage at the two chosen sites, the 'leads' being numbered as follows:

(i) Lead 1: right arm and left arm;
(ii) Lead 2: right arm and left leg;
(iii) Lead 3: left arm and left leg.

The right leg is not used for recording the ECG since it is furthest from the heart, but a neutral electrode is often applied to this limb to 'earth' the subject and hence minimise interference from the mains and electrical apparatus.

(b) Unipolar leads

These keep one electrode at zero potential while the other records changes in potential at selected points. A common arrangement is to connect the three limb electrodes (arms and left leg) together to a common 'neutral' point and the active electrode is positioned on the chest over the heart in one of six specified locations. These leads are identified as V1, V2,... V6.

Three other arrangements measure the difference between the potential in one limb and the average potential in two other limbs, the classification being:

(i) lead aVR (single limb–right arm);

(ii) lead aVL (single limb–left arm);
(iii) lead aVF (single limb–left leg).

Thus, there are twelve different electrode positions for obtaining an ECG, and each position produces its own characteristic trace.

Uses of the ECG

Any deviation from the 'normal' ECG generally indicates some cardiac disorder. For example:

(a) Irregular pumping produces correspondingly irregular trace repetition.
(b) Insufficient ventricular contraction, caused possibly by scar tissue on the heart, leads to a reduced QRS pulse height.
(c) In heart failure, the interval QT is lengthened.
(d) In ventricular fibrillation, the rapid twitching of the ventricular muscles with no pumping action, gives an easily recognisable jagged trace.

The characteristic traces so produced enable such disorders to be diagnosed and possibly treated.

The ECG may also be used as a warning device in patient-monitoring systems. These are important, particularly in intensive care units and during surgery when the ECG may be displayed on a CRO continuously so that the surgical team is instantly alerted to a crisis situation. For example, in the event of ventricular fibrillation, a defibrillator is employed applying a high voltage shock which causes all the cardiac muscle fibres to contract simultaneously. The rhythmic contractions then recover either spontaneously or with the help of a pacemaker.

The advent of miniaturisation has extended the use of the ECG. A small portable instrument, incorporating a tiny CRO, may be carried by a doctor to a patient's home or to the scene of an emergency. Further, the development of telemetry systems (see Chapter 8) enables patients to be monitored and yet remain mobile.

Electroencephalography

Brain activity

The millions of nerve cells in the brain are connected in intricate patterns and electrical activity in one group of cells can trigger activity in another group, which then triggers another group, and so on. This kind of activity continues incessantly giving rise to complex 'brain waves'. These are basically due to changes in the biopotentials across the nerve cell membranes.

The electroencephalogram (EEG)

The electrical activity within the brain may be detected by electrodes place on the skull. The voltages received, typically about $50\,\mu V$, are amplified and recorded either using a pen chart recorder or a CRO. The resulting trace is called an electroencephalogram (EEG).

The waveform contains many variations of frequency and amplitude, and although a repetitive pattern is not distinguishable as in the ECG, certain frequency ranges or 'rhythms' are detectable. The alpha rhythm contains frequencies between about 8–13 Hz, and seems to be related to mental alertness and concentration. It is most prominent when the eyes are closed and the mind relaxed, and a decrease in amplitude occurs when the brain is concentrating. Other ranges include the beta range (about 14–100 Hz), the delta range (between 0.2–3.5 Hz) and the theta range (from 4–7 Hz).

EEG electrodes are similar to, but smaller than, ECG electrodes. Various locations on the scalp can be used, depending on the region of study, and needle electrodes can be employed for greater accuracy and to reduce artifacts. Since most EEG signals contain only low frequencies, a low-pass filter can be used to remove any high-frequency noise, for example due to muscular activity.

Uses of the EEG

Brain disorders or damage are indicated by abnormal EEG patterns. For example, tumours (depriving cells of their blood supply and making them pulse more slowly) may be diagnosed and located; epilepsy is easily recognised by the characteristic spike-type EEG it produces. In addition, the EEG may be used to monitor the effects of medication or the depth of anaesthesia in the operating room, by studying the alpha rhythm.

Exercise 5

1 What is meant by an action potential and how can it lead to the production of an ECG?

Discuss briefly how disorders of the heart may be diagnosed from an ECG. What are artifacts, and how might they be minimised?

2 How good an electrical conductor is the body, and how is it protected from electrical shocks? Is the body more prone to electrical shock when it is dry or wet? Give reasons.

Discuss any problems which might arise when making electrical contact between the body and instruments such as an ECG machine, and suggest possible remedies.

Describe how ECG electrode sites are chosen and prepared.

3 (a)   (i) What are the special properties required of an amplifier employed in an electrocardiograph? (Circuit details are not required.)

(ii) State the precautions which need to be taken to ensure that a good signal voltage is transferred from the body to the amplifier.

(iii) Why should the patient be as relaxed as possible in order to obtain a good electrocardiogram?

(b) Sketch a graph of potential difference, obtained at the electrodes, against time for the cardiac waveform obtained from a single beat of the heart in a healthy person. Give approximate scales on the axes and label your diagram with the chief features of interest in cardiology   [JMB]

4 Explain the origin of 'brain waves' and describe how they may be displayed. What are the major uses of the EEG?

5 (a) Name **two** types of bioelectric signals which are often measured at the surface of the body. Sketch the waveform you would expect to observe for **one** of these signals, indicating the type it represents.

State, giving the reason, **one** precaution you would take in attaching electrodes to the surface of the skin to obtain satisfactory signals.

The signal at the electrodes is often very small and requires considerable amplification. State **two** further requirements, apart from large amplification, which an amplifier should fulfil if it is to be suitable for such measurements, explaining why they are important. (No electrical circuit diagrams of amplifiers are required).

(b) Describe, with the aid of a diagram if appropriate, **one** type of pressure measuring instrument or pressure transducer, mentioning how it is connected to the source of pressure. Give **one** application in medicine for which you consider it suitable.   [JMB]

6 | Temperature measurement

Body temperature and its limits

Under normal conditions the body maintains the central organs such as the heart, lungs, abdominal organs and brain, at a fairly constant temperature of about 37°C, known as the 'core' or simply 'body' temperature. Elsewhere, the temperature varies from location to location, being predictably lower at the skin which has an average temperature of about 33°C.

A lowering of body temperature is known as hypothermia and can be induced by exposure to cold and damp conditions. Table 6.1 summarises typical body behaviour during various stages of hypothermia.

Table 6.1   Low temperature behaviour

Core temperature (°C)	Condition of subject
37	Normal
34	Shivering, constriction of peripheral blood vessels, reduced heart rate.
33	Nervous functions depressed. Amnesia.
30	Temperature regulating system fails. Sleepiness.
28	Loss of consciousness. Respiration depressed. Cardiac fibrillation.
≈26–28	Death.

A more gradual deterioration in body function is observed during exposure to high temperatures caused, for example, by hot surroundings, heat therapy, fever or vigorous exercise. The condition, known as hyperthermia, is characterised by dilation of peripheral blood vessels, increased heart rate and cardiac output, and reduced blood flow to the brain which may result in unconsciousness. At about 41°C, the central nervous system starts to deteriorate, convulsions begin, and finally death occurs between 43°C and 45°C, when protein denaturation commences.

Heat losses from the body

Heat transfer mechanisms

Heat may be lost from the body by means of radiation, conduction, convection, evaporation, respiration and excretion. Table 6.2 shows the relative contribution from the different mechanisms (as a percentage of the total rate of heat loss) in various situations.

Table 6.2   Heat loss from the body

Activity	Body's rate of heat loss (W)	% of skin covered	% rate of heat loss by: Radiation	Conduction and convection	Evaporation	Respiration and excretion
Studying, fully clothed (21°C)	175	85	20	68	10	2
Running mile race (16°C)	1750	25	20	20	50	10
Sunbathing (32°C)	400	15	8	10	80	2
Walking, heavily clothed (−18°C)	400	95	8	50	2	40

Radiation

The manner in which an object absorbs and radiates energy depends upon the nature of its surface. For a given wavelength, a perfect absorber, which is also a perfect radiator, absorbs all radiation incident on it and is known as a black body.

The spectrum of wavelengths emitted by a black body radiator depends on its temperature (see Fig. 6.1). The higher the temperature, the greater the proportion

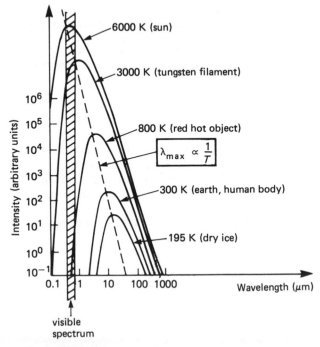

Figure 6.1   Black body radiation at various temperatures

of short wavelength radiation. For example, the sun's radiation ($T = 6000$ K) reaches a peak at about 0.5 μm (500 nm in the visible green which conveniently corresponds to the spectral region of maximum sensitivity of the eye); a red hot object ($T = 800$ K) has its maximum at about 4 μm (in the near infra-red); and the human body ($T = 300$ K) radiates maximally at 9.5 μm (also in the near infra-red). Such behaviour is summarised in Wien's law, which states that the wavelength at which maximum emission occurs, λ_{max}, is inversely proportional to the absolute temperature of the black body:

$$\lambda_{max} \propto \frac{1}{T}$$

A perfect black body radiates energy according to Stefan's law, which states:

$$E = \sigma T^4$$

where E is the total energy radiated per unit area per second by a black body at a temperature of TK, and σ is a constant known as Stefan's constant:

$$\sigma = 5.7 \times 10^{-8} \text{ W m}^{-2} \text{ K}^{-4}$$

If the body is not a perfect black body but only radiates a certain fraction e of the energy per unit area per second radiated by a perfect black body at the same temperature then:

$$E = e \sigma T^4$$

and e is known as the emissivity of the radiator. In the wavelength range in which it radiates, the human skin, regardless of its colour, has an emissivity of very nearly one, and thus radiates essentially as a perfect black body.

If the skin temperature is T_sK and the body is in an environment of temperature TK, then the net rate at which energy is removed per unit area by radiation is:

$$E = \sigma T_s^4 - \sigma T^4$$
$$= \sigma(T_s^4 - T^4)$$

If the effective radiating area of the body is A m^2, then the net heat lost per second from a body by radiation is:

$$H_R = \sigma A(T_s^4 - T^4) \qquad \text{W} \qquad [6.1]$$

The effective radiating area A is less than the true body surface area and depends on posture, position and the degree of clothing. For an unclothed subject, A may be between about 70–85 per cent of the body's true surface area, which is about 1.8 m^2. Thus, in an ambient temperature of 295 K (22°C) an unclothed subject will radiate about 100 W.

The human body only behaves like a black body in a restricted wavelength range corresponding to its own emissions and if radiation outside this range falls on the body an appreciable fraction of it will be reflected. Such is the case with radiation from the sun and the exact proportions which are reflected and absorbed depend on many factors including wavelength and skin colour.

Conduction and convection

The rate H_D at which heat is conducted through a medium of thermal conductivity k is given by:

$$H_D \text{ per unit area} = k \times \text{temperature gradient}$$

$$\frac{H_D}{A} = k \frac{(T_2 - T_1)}{d}$$

Heat conduction from the body core to the skin and thence to the surroundings is inhibited by the low values of k for body materials (see Table 6.3).

Table 6.3   Average thermal conductivities

Medium k (W m^{-1} K^{-1})	copper 385	water 0.6	fat 0.046	skin 0.042	wool 0.04	air 0.025

The presence of subcutaneous fat, hair and clothing all tend to reduce H_D.

Convection losses are either natural (due to rising air currents around the warm body) or forced (induced, for example, by the wind or a fan). The rate of heat loss by convection, H_C is clearly influenced by the subject's position (sitting, standing, lying), his movement, and the environment, but an approximate relationship is given by:

$$H_C = cf(v)(T_s - T)$$

where c is a constant and $f(v)$ is a function related to 'wind chill' which increases rapidly with wind velocity.

Evaporation

When the ambient temperature approaches or even exceeds skin temperature, radiation, conduction and convection losses are considerably reduced since they all rely on a positive difference in temperature between the skin and the surroundings. Heat loss by evaporation, (through sweating or perspiring) then becomes the dominant mechanism of heat transfer.

Evaporation takes place at the skin surface, the water changing into a saturated vapour at about skin temperature and then expanding to an unsaturated vapour in the surrounding air. Although a small proportion of the total heat provided is used to accomplish the latter expansion, most of the heat is used in the vaporisation process. Thus:

$$H_V \approx mL$$

where H_V is the rate at which heat is lost by the evaporation of m kg per second of water of specific latent heat of evaporation L J kg^{-1}.

H_V depends on several factors, including exposure, air and skin temperatures, humidity and air circulation. Under favourable conditions, the eccrine glands can be stimulated to sweat up to 1 kg per hour. Since L (at 33°C) is 2425 kJ kg^{-1}, this corresponds to a rate of heat loss of 2425 kJ h^{-1} ($\approx$ 675 W).

Respiration and excretion

In addition to the heat conducted out through the skin, some heat is lost indirectly by conduction in the processes of respiration and excretion. When air is inspired, it is heated by conduction when it comes into contact with the warm nasal cavities. Similarly, food at a temperature lower than that of the mouth extracts heat by conduction from the body to warm the food to body temperature. Subsequently, these quantities of heat are lost from the body in the processes of expiration and excretion.

Moisture is lost from the lungs during respiration: expired air is normally about 95 per cent saturated at temperatures of 33–35°C. A resting man under comfortable conditions loses about $0.03\,kg\,h^{-1}$ of water via respiration, equivalent to a rate of loss of energy of about 20 W. Respiration losses are particularly important when the surrounding air is dry and/or cold.

Control of body temperature

Heat is produced in the body by metabolism and physical activity. In order to maintain a steady state (i.e. to keep the core temperature constant) heat must be removed from the body at a rate equal to that at which it is produced. To achieve this control of body temperature, the body:

(a) detects changes in core temperature and initiates stabilising action;
(b) may change the rate of heat production either involuntarily, (varied metabolic activity, shivering) or voluntarily (stamping feet, clapping hands);
(c) may vary the rate of heat removal by such means as:
   (i) curling up in cold weather or spreading out in warm weather to vary the effective body surface area;
   (ii) vasoconstriction (reduction in diameter of peripheral blood vessels) in cold conditions, or vasodilation (increase in blood vessel diameter) in warm conditions, so controlling blood flow rate and heat losses predominantly by conduction and radiation;
   (iii) roughening of skin (goose-pimples) and erecting hairs in cold conditions to reduce air currents close to the skin thus decreasing convective losses;
   (iv) wearing suitable clothing (e.g. smooth white clothes in hot climates, thick woollen clothes in cold climates).

Temperature transducers

A transducer is a device for translating one type of varying signal to another. For example, the rods and cones of the eye are optical transducers in that they convert optical signals into the electrical signals conveyed along the optic nerves; a microphone is an acoustic transducer, changing acoustic inputs into electrical signals; a thermometer is a temperature transducer, translating temperature changes into, for example, volume changes of mercury.

The most versatile of the temperature transducers are those with electrical outputs. Not only can they be made small and sensitive, but they also permit the remote recording of signals from possibly inaccessible locations. Those most frequently used for medical applications are based on the thermoresistive (variation of electrical resistance with temperature) and thermoelectric (variation of thermoelectric e.m.f. with temperature) effects.

In selecting the most suitable thermometer for a particular temperature measurement, several factors need to be considered: the size and thermal mass of the 'probe', since these determine the disturbance imposed by the measurement and the speed of response; ease and reproducibility of measurement; range and sensitivity of the instrument; and, possibly of most importance, the location of the measurement.

Site of temperature measurement

An estimate of core temperature may be made most simply using a mercury-in-glass clinical thermometer in the mouth or armpit. The thermometer must be left in position for a minute or two to reach body temperature.

A more accurate assessment of core temperature is obtained by measuring the temperature in the rectum at a depth of 8 cm or more. A rectal thermocouple, consisting of a thin metal cap housing the thermocouple junction cemented on to a flexible thermoplastic tube, provides good thermal contact and is the preferred instrument for this location.

Internal temperature monitoring, for example in the heart, blood stream, respiratory tract and tissue, demands miniature temperature transducers, often of the thermoresistive type.

Skin temperature measurement requires a 'probe' of negligible thermal capacity which is approximately achieved using thermoresistive or thermoelectric transducers. However, any 'contact' instrument insulates the skin to a certain extent and for greater accuracy the radiation thermopile is used.

Types of thermometer

The mercury-in-glass clinical thermometer

When the bulb of the thermometer (Fig. 6.2) is put into the mouth or armpit, the expanding mercury forces its way past a fine constriction to form an opaque 'thread' in the capillary tube. When the bulb is removed, this constriction holds the thread in place for subsequent observation. The thermometer is reset by shaking.

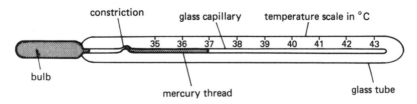

Figure 6.2   The mercury-in-glass clinical thermometer

The sensitivity of the instrument is increased by using a large bulb and narrow bore capillary, thus giving rise to relatively large changes in the length of the mercury thread for small temperature changes. The accuracy with which it can be read is increased by shaping the front of the glass tube like a lens, to magnify the thread and scale. The accuracy of measurement achievable, however, is limited by:

(a) lack of good thermal contact with the object of measurement;
(b) alteration of the measured temperature due to the large thermal capacity of the bulb.

Thermoresistive thermometers

In the conventional resistance thermometer the temperature-sensitive resistor is a pure metal wire, commonly of platinum, of resistance from a few ohms to a few hundred ohms. To achieve a reasonable resistance the coil tends to be rather long, making it unsuitable for mounting in very small diameter tubes such as catheters or hypodermic needles. Platinum exhibits a wide linear resistance–temperature relationship of the form:

$$R_T = R_0 (1 + aT)$$

where R_T and R_0 are the resistances of the wire at $T°C$ and $0°C$ respectively and a is a constant known as the temperature coefficient of resistance and given by the slope of the graph in Fig. 6.3. Since the value of a for platinum is about 0.4 per cent per degree centigrade, it is necessary to use a Wheatstone bridge circuit with a sensitive indicator to obtain accurate temperature measurements (Fig. 6.4). At balance:

$$\frac{R_A}{R_B} = \frac{R_T}{R_L}$$

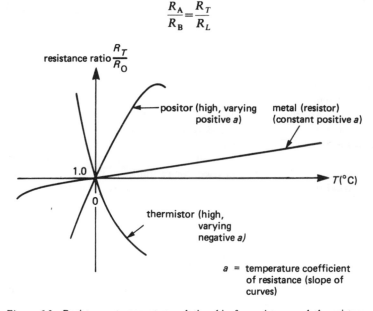

Figure 6.3 Resistance–temperature relationship for resistors and thermistors

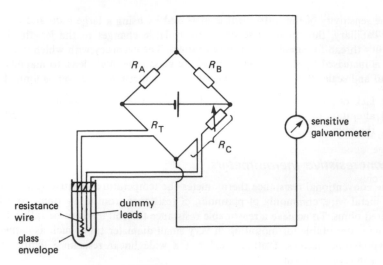

Figure 6.4 Platinum resistance thermometer

and, with the use of standard resistances and suitable calibration, R_T (and hence T) can be found.

In order to compensate for the change in resistance of the wires connecting the resistance element to the bridge, 'dummy' wires of identical resistance are connected to the opposite arm of the bridge, as shown. Using this technique, a temperature can be measured with an accuracy of a few hundredths of a degree centigrade.

Thin-film resistance thermometers, in which the element is a small platinum film of resistance about $15\,\Omega$, have been developed and these are small enough to be mounted in a catheter tip. Such instruments can be used to investigate blood flow patterns in the cardiovascular system.

Thermoresistive transducers based on the properties of certain ceramic-like semiconductors are particularly useful due to their small size, excellent long-term stability and relatively large temperature coefficient. A transducer of this type is the thermistor, a verbal compression of 'thermally-sensitive resistor'. Its temperature coefficient is negative (see Fig. 6.3), and is typically about ten times greater than that for a metal making it significantly more sensitive. The positor (Fig. 6.3) with its high positive coefficient, is similarly sensitive. Miniaturised thermistors may be mounted in catheters and used, for example, to monitor blood temperature. A typical thermistor thermometer (Fig. 6.5) consists of a bead made from a

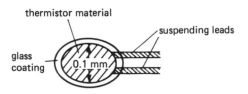

Figure 6.5 Thermistor probe

mixture of nickel, cobalt and manganese oxides, suspended from lead wires and mounted in a thin protective glass envelope. Some thermometers of this type can measure temperature differences of the order of 0.01°C. Disadvantages of thermistors include greater cost and shorter life than pure metals, and a non-linear (approximately exponential) resistance–temperature relationship, yielding non-linear scales.

In general, the current through the resistive element should be minimised to avoid self-heating errors. However, heated thermistors may be placed in the respiratory air-stream to detect respiration. The thermistor is maintained in a bridge circuit at a temperature above ambient and it is cooled once every respiratory cycle by the passage of cooler inhaled air. The thermistor resistance then changes, thus varying the bridge output voltage, which triggers a counting circuit. The frequency of the output waveform thus supplies the respiratory rate, whilst the amplitude of the waveform gives an estimate of air volume.

Thermoresistive thermometers are widely used when remote temperatures (e.g. in an intensive care unit) are to be recorded at a central nursing station.

Thermoelectric thermometers

If two unlike metals A and B are joined together (see Fig. 6.6) and their junctions kept at different temperatures, an e.m.f. is generated known as the thermoelectric e.m.f. which depends on the difference in temperature between the junctions. Such a device, known as a thermocouple, is suitable for use as a temperature transducer in which either the thermoelectric e.m.f. is measured using a sensitive potentiometer, or the thermoelectric current is monitored using an accurate galvanometer. Junctions with the external circuit must be maintained at a constant temperature to prevent spurious thermoelectric e.m.f.'s being set up here.

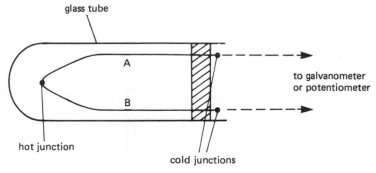

Figure 6.6   A thermocouple

The cold junction should be maintained at a constant reference temperature, commonly 0°C or room temperature. Since, however, this is often inconvenient for ward or operating room use, an automatic cold junction compensation device is frequently incorporated. The thermocouple cold junction is mounted close to the galvanometer which has one of its coil suspension springs made from a bimetal. Although changes in room temperature affect the cold junction and give rise to spurious e.m.f.s, they also cause the bimetal spring to twist in such a direction as to exactly compensate the deflection of the galvanometer produced by such spurious e.m.f.s.

Although the thermocouple has been somewhat overshadowed by the thermistor as a temperature sensor, new fabrication techniques are providing thermocouples of micrometre dimensions and fast response times (≈ 0.1 s), shown in Fig. 6.7. In addition, thermoelectric thermometers tend to be robust,

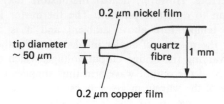

Figure 6.7   Miniature thermoelectric probe

easily reproducible, relatively cheap, and of small thermal capacity. They find application in rectal thermometers, thermocouple probes and blood flowmeters. The latter consists of two small thermocouples and a heating element in a tube which is inserted into the blood stream. The thermocouples detect the blood temperature before and after it reaches the heating element and the difference in temperature is related to the blood flow.

When several thermocouple junctions are incorporated into an instrument known as a thermopile (see Fig. 6.8) the thermoelectric e.m.f generated when the hot junctions are warmed is significantly greater than that due to one junction alone. Such a device, with its receiving cone positioned close to, (≈ 4–7 mm) but not touching, the skin, can therefore detect heat radiated from the skin. An estimate of skin temperature is then made using Stefan's law:

$$E = \sigma A T_s^4$$

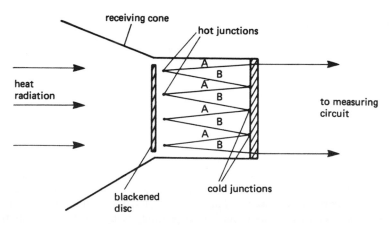

Figure 6.8   A thermopile

The instrument may be calibrated to read temperature directly.

Thermography

General principles

Using equation [6.1] comparative measurements of H_R across the body surface yield corresponding variations in T_S, the skin temperature. In the technique of thermography, the infra-red radiation emitted by a scanned body is detected and analysed to produce a 'thermal image' of the body surface. Such an image is called a thermogram and usually warm areas appear light, cool areas dark, with shades of grey for intermediate temperatures. Local disorders, such as infection, arthritic disturbances, burns, bruising, superficial tumours and general inflammation, lead to regions of increased blood supply and hence increased surface temperature, which then appear as 'hot spots' on the thermogram.

The thermographic camera

The subject lies on a table above which is mounted the scanning thermographic camena (see Fig. 6.9). Infra-red rays from a point on the subject are reflected by the plane image-scanning mirror onto a concave mirror, which focuses the radiation on to a small infra-red detector. This transduces the incoming infra-red signal into an electrical signal, which is then passed via an amplifier to a cathode ray tube for

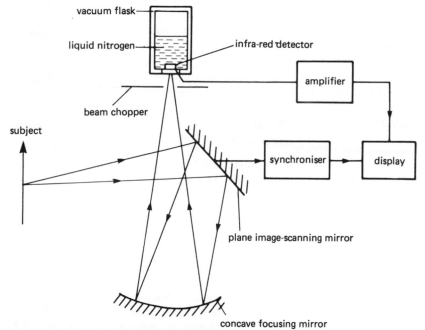

Figure 6.9   Thermographic camera

display. On the screen is produced a spot of light, whose intensity is a measure of the temperature of the point monitored. By oscillating the image-scanning mirror about a vertical axis a strip of the subject is scanned. After each sweep, the angle of the mirror is changed slightly (by rotating it through a small angle about a horizontal axis) and a new 'strip' is scanned. These movements of the mirror are fed to a synchronising unit and thence to the display so that the position of the spot on the display screen is exactly synchronised with the point of the subject whose temperature is being monitored. Thus, a picture is rapidly built up on the screen, displaying visually the surface temperature distribution of the subject. When examining a patient at a distance of about 3 m from the camera an area of about 60×100 cm can be scanned in about four minutes, and a temperature resolution of about 0.2°C is obtainable for a geometrical resolution of 3 mm on the display.

The infra-red detector is a photoconductor made from a single crystal of indium–antimony alloy. When radiation falls on the crystal, its electrical resistance decreases by an amount determined by the intensity of the radiation. The resistance change is translated into a voltage variation for subsequent amplification and display. This voltage variation thus reflects the original temperature distribution being monitored. Since the incident infra-red signals are of very low intensity, background radiation or noise in the detector can be troublesome. Therefore, the detector is cooled to liquid nitrogen temperature (77 K) to reduce such thermal noise.

In order to compensate for electronic drift, the incident radiation is 'chopped' by a polished rotating blade, so that the detector is alternately presented with images of itself and the subject. This provides an inherent reference temperature (that of the detector) against which that of the subject is compared.

Germanium and silicon lens-focusing elements are used to double as filters, excluding spurious short-wavelength radiations, whilst transmitting the wavelength range emitted by the human body.

To improve the accuracy, reproducibility and clarity of thermograms, the examination room should be cool, dry and under thermostatic control, and a ten-minute pre-cooling of the uncovered skin at about 20°C is recommended.

Applications of thermography

(a) Detection of tumours
Any tumorous growth tends to draw more blood than the surrounding normal tissue. This extra blood leads to local 'hot spots' which are seen on the thermogram as light areas. Thermography has proved useful in diagnosing, for example, breast and thyroid tumours.

(b) Mapping of blood vessels
The section of the body to be studied is first cooled to sub-normal temperature, and a thermogram is taken soon after removal of the cooling agent. Since blood vessels heat up faster than the surrounding tissue, they show up as light areas on the thermogram, which may therefore be used to locate constrictions or abnormal dilations, as well as to monitor the effects on blood vessel condition of drugs or diet. Even without elaborate precooling procedures, impaired circulation may be seen, often considerably in advance of detection by clinical means.

(c) Investigation of bone fractures

Since these cause irritation of the surrounding muscles and tissues, hot spots on a thermogram may indicate a hair line fracture undetectable using X-rays. The extent and location of arthritic disturbances may similarly be estimated.

(d) Gangrene

A thermogram of the foot indicates the extent and seriousness of the disease, thus facilitating recommendations for treatment.

(e) Placental localisation

During pregnancy, the location of the placenta can be determined accurately and without any danger to mother and child. This knowledge is often vital for a safe delivery.

(f) Burns and frostbite

The depth of tissue destruction may be estimated and precisely mapped, for removal, if necessary, by surgery.

Exercise 6

1 Discuss the major mechanisms responsible for the removal of heat from the body. What would you consider their relative importance to be in the case of:
   (a) a person in a sauna;
   (b) a heavily-clothed hiker on a cold, dry day;
   (c) a sun-bather on a hot, dry day?

2 Explain carefully how man can live in environments of different temperatures ranging from about $-30°C$ to $40°C$ whilst maintaining his deep body temperature reasonably constant at about $37 \pm 2°C$.
   In one hour, a subject loses 0.04 kg of water from his skin and lungs. If this loss represents 20 per cent of his total body heat loss, find the amount of heat he loses per second. (Latent heat of evaporation of water at average body temperature $= 2.4 \times 10^6$ J kg^{-1}.)

3 Describe how an accurate measurement of skin temperature can be made.
   At the beginning of a race, an athlete loses heat at a rate of 1400 W. However, his body is generating heat at a rate of 1750 W. If his thermal capacity is 2.4×10^5 J K^{-1}, find the initial rate at which his temperature starts to rise.
   If his heat production and loss rate remain at these values, what would his temperature rise be at the end of a one-hour race? Does this happen in practice? Discuss.

4 Explain the principles of operation of a thermographic camera. What factors contribute to the clarity of the final image?
   During a medical examination, an unclothed subject loses heat at the rate of 85 W; 70 per cent of this loss is by radiation, and the ambient temperature is 25°C. If the effective radiating area of his skin is 80 per cent of his true body surface area, which is 1.8 m^2, find his average skin temperature. State any assumptions you make. (Stefan's constant $= 5.7 \times 10^{-8}$ W m^{-2} K^{-4}.)

5 Select, giving reasons for your choice, a thermometer suitable for:
   (a) measuring core temperature during hypothermia (the temperature of the

oesophagus is commonly taken as a measure of cardiac and cerebral temperature);

(b) monitoring temperature in an intensive care unit;

(c) the routine temperature measurements of convalescing patients in hospital;

(d) measuring skin temperature.

For an unclothed subject, the effective radiating surface area is 85 per cent of his true body surface area, the latter being 1.8 m². Find the net rate at which his body loses heat by radiation when he is in an ambient temperature of:

(e) 32°C, when his average skin temperature is 35°C, e.g. sun-bathing;

(f) 21°C when his average skin temperature is 33°C; e.g. medical examination. (Stefan's constant $= 5.7 \times 10^{-8} \ Wm^{-2} \ K^{-4}$.)

6 (a) (i) State what kind of thermometer you would use to measure the difference in temperature which occurs between inhaled and exhaled air during respiration. Give reasons for your choice.

(ii) Explain briefly why body temperature rises during vigorous exercise, even though the heat lost from the body due to respiration increases.

(b) (i) A person inhales in one breath $5 \times 10^{-4} \ m^3$ of dry air at atmospheric pressure and 20°C. The air is then warmed to the body core temperature of 37°C in the lungs. If the person takes twelve breaths per minute, calculate the heat transferred per minute to the air from the body. Assume that there are no pressure changes in the inhaled air during respiration.

(ii) During each breath, $2.2 \times 10^{-5} \ kg$ of water vapour is formed in the lungs and is then expelled with the exhaled air. If the specific latent heat of evaporation of water at 37°C is $2.3 \times 10^6 \ J \ kg^{-1}$, calculate the heat lost per minute due to this process.

Comment on your result. (Density of dry air at 20°C and atmospheric pressure $= 1.2 \ kg \ m^{-3}$; specific heat capacity of dry air at atmospheric pressure $= 1.0 \times 10^3 \ J \ kg^{-1} \ K^{-1}$.) [JMB]

7 | Pressure measurement

Electrical pressure transducers

Transduction process

The pressure to be measured is generally applied to one side of a deflectable diaphragm. The consequent transduction from diaphragm motion to electrical signal may occur using:

(a) resistance (variable resistance transducers or strain gauges);
(b) capacitance (variable capacitance transducers);
(c) inductance (variable inductance transducers); or
(d) other intermediate properties.

External transducers

If the pressure transducer to be used is outside the examined structure, a hydraulic system is needed to transmit the pressure from the site of measurement to the pressure transducer. Such an arrangement is found in the diaphragm manometer shown in Fig. 7.1 in which the pressure to be measured is applied to one side of a thin but stiff diaphragm (of phosphor bronze, stainless steel or beryllium copper) which is attached firmly by its edge to the wall of the diaphragm chamber. The resulting deflection of the diaphragm is then transformed into a corresponding

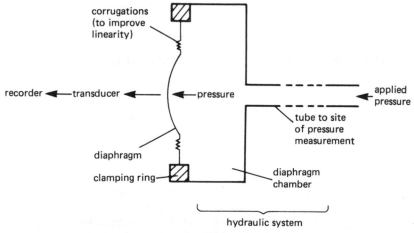

Figure 7.1 Diaphragm manometer

electrical signal by one of the many types of electrical pressure transducers. The final electrical signal, fed to a suitable recorder, gives a measure of the unknown pressure.

The complete instrument (hydraulic system, diaphragm, transducer and recorder) shows a relatively poor frequency response due to the inertia of the hydraulic system which cannot follow high frequencies as well as low frequencies. Also, it displays resonance and if the fluctuating pressure being monitored contains frequencies near the system's resonant frequency, these predominate and make the interpretation of results difficult. In practice, its use is limited to sites near the transducer and pressures which do not have high frequency fluctuations.

Internal transducers

When small internal pressure transducers are inserted at the site of measurement (see Fig. 7.2) the transduction to electrical signals is achieved *in situ*. These signals are then relayed to an external recording system either directly via connecting wires along a catheter (a long plastic tube for inserting into body tracts), or indirectly using a radio transmitter (see Chapter 8).

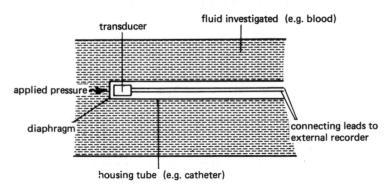

Figure 7.2   Internal pressure transducer

Variable resistance transducers (strain gauges)

Resistance change and its measurement

A resistance element (the transducer) is attached to the deflectable diaphragm so that its axis lies along a radius of the diaphragm. As the diaphragm distorts under the application of the unknown pressure, the resistance element is either compressed or extended (as in Fig. 7.3(a)), depending on its location. This results in a change of element resistance (since $R = (\rho l)/a$) which is then measured and interpreted as a pressure reading. The resistance element is known as a strain gauge, since it records the strain in the associated diaphragm.

In order to compensate for any resistance changes occurring due to temperature changes, a full bridge circuit (Fig. 7.3(b)) is commonly employed to measure

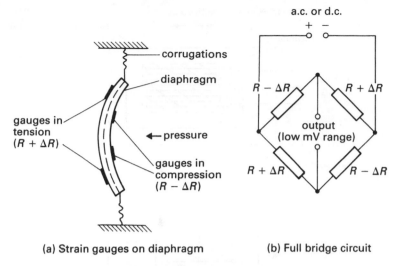

(a) Strain gauges on diaphragm   (b) Full bridge circuit

Figure 7.3   Measurement of gauge resistance changes

resistance. Two pairs of identical strain gauges, positioned on the diaphragm so as to undergo strains in opposite directions, are connected in the bridge circuit as shown. Since both arms of the bridge are equally affected by temperature changes, the imbalance signal gives a true record of ΔR due to strain alone. Suitable calibration then yields pressure directly.

Silicon, being very sensitive to strain, has been widely used as a gauge material although its significant temperature coefficient of resistance makes it unsuitable for use where temperature fluctuations are large.

Bonded and unbonded strain gauges

In the basic bonded gauge, shown in Fig. 7.4(a), the resistance element is formed into a zig-zag pattern and firmly bonded to a backing medium such as paper, cloth or plastic. The more sophisticated, and very successful, silicon-bonded strain gauge consists of a single crystal element (about 0.01 mm thick and 0.25 mm wide) mounted in a substrate of epoxy resin and glass fibre, with nickel strips for electrical connections. The complete assembly is coated with epoxy resin for protection.

A typical unbonded strain gauge (Fig. 7.4(b)) consists of four strain-sensitive wires (A, B, C and D) connected between two frames, the inner movable one fitting loosely inside the fixed one. The inner frame is attached to the diaphragm to which the unknown pressure is applied, and any movement of this frame produces opposite strains in A and B, and in C and D. Once again, a full bridge circuit can be used to measure resistance changes and thus pressure. Outputs are small (≈ 50 mV per mmHg) and therefore a high-gain amplifier is needed to follow the bridge.

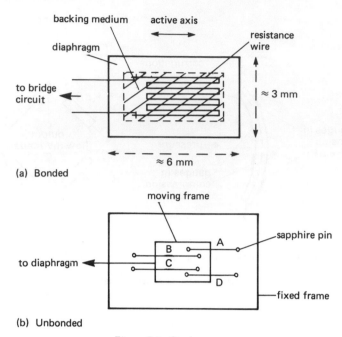

(a) Bonded

(b) Unbonded

Figure 7.4   Strain gauges

Variable capacitance transducers

Single capacitance transducer

The capacitance C of a parallel-plate capacitor is given by:

$$C = \frac{\varepsilon A}{d}$$

where ε is the permittivity of the medium between the plates, A is the cross-sectional area of one plate and d is the distance between the plates, which is small compared with the area of the plates. By moving either one of the plates, or the dielectric, as shown in Fig. 7.5, the capacitance changes by an amount dependent on the displacement. Hence, by coupling the capacitor to a diaphragm to which pressure is applied, any movement of the diaphragm is monitored by a corresponding change of capacitance. In some cases the diaphragm itself, made of metal, forms one plate of the capacitor and is located only a few hundredths of a millimetre from the other fixed plate.

The various methods for measuring the changing capacitance usually employ an alternating current. For example, the transducer can form one arm of an a.c. bridge circuit the imbalance of which gives a measure of the change in capacitance, and hence pressure. Alternatively, the transducer may be included in a simple tuned circuit such that capacitance changes produce frequency changes, which are subsequently interpreted as pressure changes. Perhaps the simplest arrangement is

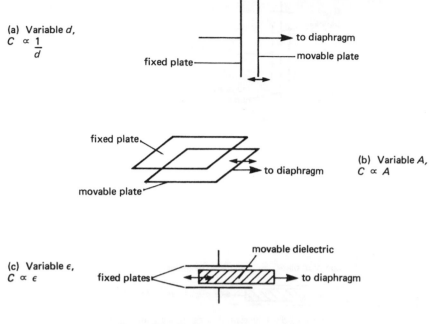

(a) Variable d,
$C \propto \dfrac{1}{d}$

to diaphragm
movable plate
fixed plate

fixed plate
to diaphragm
movable plate

(b) Variable A,
$C \propto A$

(c) Variable ϵ,
$C \propto \epsilon$

movable dielectric
fixed plates
to diaphragm

Figure 7.5   Single capacitance transducers

the R–C series circuit in which the output V_c across the variable capacitor gives a measure of C.

Differential capacitance transducer

As with the strain gauges, compensation for external factors, such as temperature, is achieved by using two identical capacitors in opposite arms of an a.c. bridge circuit, so that one capacitance increases $(C+\Delta C)$ whilst the other decreases $(C-\Delta C)$ when pressure is applied. The devices designed for this purpose are called differential capacitance transducers (see Fig. 7.6).

Variable capacitance transducers can measure both large and small pressures, have high sensitivities, and display rapid response times. However, the stray capacitance from coaxial cable connections can reduce sensitivity and give rise to spurious signals upon movement of the cables.

Variable inductance transducers

Single inductance transducer

The self-inductance L of a coil depends partly on the amount of high-permeability core inside the coil. If such a core is attached to a diaphragm to which an

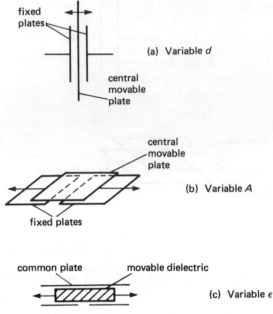

(a) Variable *d*

(b) Variable *A*

(c) Variable *ε*

Figure 7.6   Differential capacitance transducers

unknown pressure is applied, any deflection of the diaphragm produces a change in *L*. An associated circuit (for example, a simple oscillator, a series circuit, or an a.c. bridge) can monitor changes in *L* and hence in applied pressure.

Linear variable differential transformer (LVDT)

A primary coil (see Fig. 7.7) is energised using an audio-frequency alternating voltage. This induces voltages in two identical secondary coils located on either side of the primary, and connected in series opposition so that their induced voltages act in opposition.

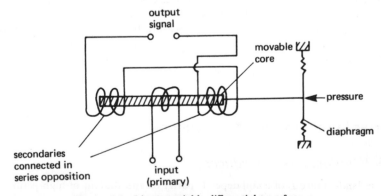

Figure 7.7   Linear variable differential transformer

The coils are wound around a movable, high-permeability, ferromagnetic core which is connected to the diaphragm to which pressure is applied. In the 'zero position' the two secondary voltages exactly cancel to give a zero output signal. When the core is displaced due to an applied pressure, one secondary voltage increases whilst the other decreases, giving an output voltage which varies linearly with core position and hence pressure.

An output of the same magnitude is obtained if the core is displaced an equal amount in either direction from its central position. To avoid any confusion of readings, the core in a directionally-sensitive LVDT is offset from the centre in its 'zero position', thus giving a finite output signal for zero applied pressure. Movement of the core in opposite directions then either increases or decreases the output which is still linearly related to the applied pressure.

The LVDT is relatively rugged, fairly insensitive to temperative changes, has good sensitivity and a rapid response time. However, at the 'null' position of the core there remains a small residual voltage which is about 1 per cent of the maximum output voltage.

Piezoelectric pressure transducers

Of the many remaining electrical pressure transducers, perhaps the most common is the piezoelectric transducer. Certain crystals produce a surface potential difference when they are compressed or extended in particular directions (see page 118). This phenomenon, known as the piezoelectric effect, thus provides a simple means of transducing strain (and therefore pressure) into electrical signals. For example, certain piezoelectric ceramics can be moulded and polarised to be pressure-sensitive in a given direction, and can yield large outputs (1 atm pressure acting on a 1 mm thick specimen can produce a voltage of 10 mV).

Because of the direct production of a signal without the use of external power supplies, such transducers are referred to as active transducers. In contrast, the transducers relying on changes in R, L and C, use supplementary power sources and are called passive transducers.

The measurement of body pressures

When selecting a suitable instrument for measuring a body pressure it is important to consider:

(a) the site of measurement;
(b) the magnitude of the pressure;
(c) the fluctuations in the pressure.

For example, venous blood pressure is low (≈ 0–10 mmHg) and fluctuations are small and slow so that an instrument with a frequency response of a few hertz. is adequate for venous blood pressure measurements. In contrast, changes in arterial blood pressure (≈ 80–120 mmHg) are rapid and demand a device with a good frequency response up to high frequencies.

Deep body sites require the use of a miniature pressure transducer situated at the tip of a catheter, whilst for locations nearer the body surface an external transducer may be used in conjunction with a short catheter or even a hypodermic needle.

Although many body pressures need to be measured (e.g. within the bladder, uterus, gastrointestinal tract, spine) by far the commonest pressure measurement is that of blood pressure.

Blood pressure measurement

Blood pressure

Conventionally, blood pressure is quoted as the pressure of blood at the height of the heart relative to atmospheric pressure. Thus, the site of pressure measurement is important and a correction of about 8 mmHg must be applied for every 10 cm that the site of measurement is below the height of the heart.

The pressure of blood in the arteries reflects cardiac activity much more than does the smaller blood pressure in the veins and the former is thus the accepted 'blood pressure'. Normally, it varies between 80 mmHg (diastolic blood pressure) and 120 mmHg (systolic blood pressure) in which case the blood pressure is quoted as 120/80.

Indirect methods of blood pressure measurement

In such methods there is no direct contact with the blood stream and hence no arterial puncture is necessary. This is a considerable advantage and makes the indirect method the preferred one for the routine monitoring of arterial blood pressure in comparatively fit patients.

The instrument used is called a sphygmomanometer and consists of a cuff arrangement connected to a mercury manometer (see Fig. 7.8). A rubber bag enclosed in a cloth cuff, is wrapped around the upper arm (at about heart level) and inflated using a small hand pump to about 200 mmHg, well above the systolic pressure. The brachial artery, the large artery in the arm, is now obstructed and silence is heard in a stethoscope placed over the artery below the elbow. The cuff pressure is then slowly reduced whilst the observer simultaneously watches the manometer reading and listens to arterial sounds through the stethoscope. As the systolic pressure is reached, a tapping sound in step with the heartbeat is heard and corresponds to the artery wall beginning to be forced open. As the cuff pressure is further reduced, the wall opens more and more and the tapping gets louder. Eventually the sounds start to get muffled and gradually fade away as the diastolic pressure is reached and the walls remain permanently open. This sequence of sounds is known as the Korotkoff sounds and the accuracy with which these can be detected and a moving mercury column can be read is about ±5 mmHg. The sphygmomanometer is thus not as accurate as electric pressure transducers, but it is simple and convenient for everyday use.

Direct methods of blood pressure measurement

Direct contact with the blood stream potentially gives a greater accuracy in blood pressure measurement and on average the systolic pressures recorded directly are about 10 mmHg higher than those recorded indirectly.

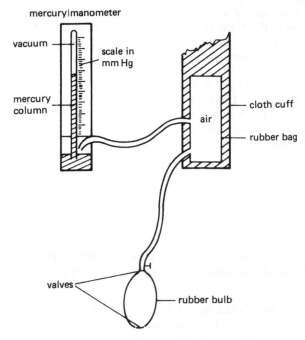

Figure 7.8 The sphygmomanometer

For measuring an average pressure perhaps the most reliable and inexpensive instrument to use is the mercury manometer (see Fig. 7.9). A simple catheter with side holes at the tip, or even a hollow needle, is inserted into the blood vessel and the pressure is transmitted to the mercury manometer via saline containing heparin to prevent blood clotting. To stop blood entering the recording system, approximately the correct pressure must be applied to the system from a press-

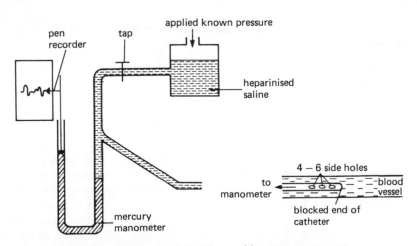

Figure 7.9 Pressure-sensing catheter with mercury manometer

urised saline reservoir via a tap. A float and pointer writing on smoked paper provide a record of the blood pressure. Because of the high inertia of the mercury and associated saline columns, the instrument cannot follow rapid pressure changes, and simply records an average value.

When fluctuating pressures are to be measured, a diaphragm pressure transducer (see above) is used which is usually mounted at the tip of a catheter and inserted where required. The catheters used, for instance, in cardiac studies are typically made of radio-opaque plastic tubing so that they are clearly visible on an X-ray, and X-ray closed-circuit television systems (see page 109) may be used to assist in their manoeuvring. The catheters are about 1–2 mm in diameter and about 80–125 cm in length.

Exercise 7

1 Compare the use of the pressure-sensing catheter and mercury manometer with that of the sphygmomanometer for the measurement of blood pressure. How important is positioning of the manometer system in each case?

What features are important in the design of a cardiac catheter tip pressure transducer?

2 State, giving reasons, which instrument is most suitable for:
(a) monitoring the arterial blood pressure of a patient in an intensive care unit;
(b) the routine measurement of arterial blood pressure during pregnancy;
(c) the measurement of venous blood pressure;
(d) respiratory pressure studies (several mmHg).
Describe in detail the functioning of one of them.

3 Describe with the aid of a diagram the structure of the linear variable differential transformer. Explain in detail the changes which occur when pressure is applied to such a transducer and how the output provides a measure of this pressure. Outline one method by which the output is monitored.

4 How does the site of measurement affect the choice of a suitable pressure-measuring instrument in medicine?

Describe briefly three different instruments for measuring body pressures and indicate for each a typical site of measurement.

5 Describe the use of strain gauges in the measurement of body pressures. How are temperature changes likely to affect results?

The gauge factor G of a steel strain gauge, defined as:

$$G = \frac{\text{fractional change in resistance}}{\text{fractional change in length}}$$

is found to be 2. It is also found that the maximum load which can be placed on a steel wire of diameter 1 mm is 155 N. If Young's modulus for steel is $2 \times 10^{11} \text{ N m}^{-2}$, find the maximum percentage change in resistance which can be measured by the gauge.

6 (a) Describe a typical transducer for pressure measurement based on change of electrical resistance, explaining its mode of action.

A resistance transducer has diameter 1 cm and is to be used to measure blood pressure in an artery. Describe the technique you would expect to be used for such a measurement.

It is proposed to make a miniature version of the transducer mounted on a catheter tip. If the same materials are used and each linear dimension reduced to one-tenth of its original value, by what factor would the resistance of the sensing element be changed?

(b) For a person of height 2.00 m, the blood leaving the heart has a mean pressure of 13.30 kPa. If the heart is twice as far from the feet as from the head, calculate the blood pressures in the lower part of the foot and the upper part of the head when the person is erect. Assume gravitational forces alone are responsible for the differences in pressure. (Density of blood $= 1.04 \times 10^3 \, \text{kg m}^{-3}$; $g = 10 \, \text{m s}^{-2}$.) [JMB]

8 | Radiotelemetry

Biomedical radiotelemetry

Telemetry is the recording of electrical quantities at a distance. In biomedical radiotelemetry, physiological data (e.g. temperature, pressure) are translated into electrical signals using a suitable electrical transducer, and the signals are then transmitted by radio waves to a distant receiver.

Long-range transmission (macrotelemetry) requires the use of a long-wire radiator, similar to a conventional aerial, worn outside the body, and permits the recording of body functions at distances of up to several kilometres. This permits the monitoring not only of mobile patients in hospitals (e.g. ECGs, blood pressure, etc.) but also of 'normal' personnel operating in strenuous situations, such as astronauts, athletes or members of the armed forces.

Short-range transmission involves the use of a tiny device known as a radio pill, or endoradiosonde, which is placed inside the body and transmits information to a nearby external receiver.

Radio pills

The whole assembly of transducer, transmitter and power supply (see Fig. 8.1) is encapsulated, giving the appearance of a 'pill' and introduced into the body by swallowing or surgical implantation. It must clearly be small, light, and capable of withstanding exposure to body fluids.

The transmitter (see below) is essentially a coil like the primary coil of a

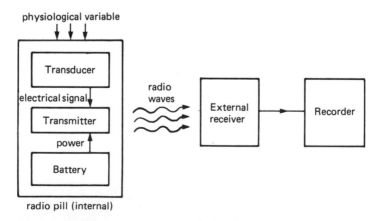

Figure 8.1   Radio pill

transformer. Some of its lines of force link a larger external coil, like the secondary of a transformer, and induce there an exactly corresponding signal which is monitored by the receiver.

The signal may become too weak to measure above background noise if the transmitter/receiver separation becomes too great, and also if the transmitter changes its orientation, since if the two coils are perpendicular to each other there is no energy coupling. This situation may arise, for example, as the pill moves along the turns of the gastrointestinal tract. Therefore, an omnidirectional system is used incorporating three mutually-perpendicular receiving coils. The output from the three coils is then either scanned cyclically or combined to give a single signal.

If the signal reaching the receiver is still too weak, the subject may carry a booster transmitter which receives the weak signal from the internal transmitter and re-radiates an amplified signal to the more distant receiver.

Pressure-sensitive radio pill

In this 'pill' (see Fig. 8.2) the deflecting diaphragm is connected to a ferrite disc[1] beyond which is a ferrite pot core containing a small coil of wire of self-inductance L. An increase in pressure reduces the air gap between lid and pot and hence increases L. The coil is included in a tuned circuit so that changes in L produce

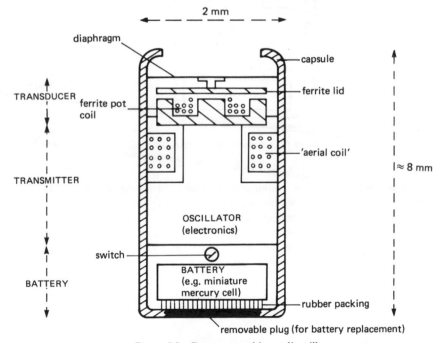

Figure 8.2   Pressure-sensitive radio pill

[1](Ferrite is a material of high permeability but low electrical conductivity).

corresponding changes in the resonant frequency. An 'aerial coil' in this tuned circuit then transmits the modulated signal to an external receiver. Frequency changes thus reflect pressure variations and the recording system is generally calibrated to register pressure directly.

Temperature-sensitive radio pill

Any temperature-sensitive radio pill should have a small thermal capacity, and a thermal conductivity and specific heat capacity approximating to that of the material in which it is situated (e.g. tissue) so that its presence does not change the thermal environment.

One of the most common types uses an inductance transducer (see Fig. 8.3). Its

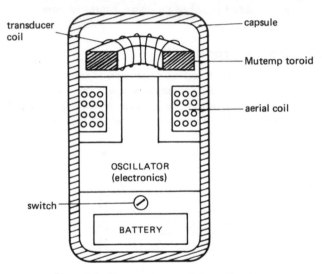

Figure 8.3   Temperature-sensitive radio pill

coil, of self-inductance L, is wound onto an insulated toroid of Mutemp HTC, a special nickel–iron alloy, whose permeability is very sensitive to temperature changes. L is also therefore temperature-dependent, displaying a variation of about 7 per cent per degree change in the temperature range of interest, (≈ 0–$40°$C).

As before, a tuned circuit is used to monitor changes in L and to transmit these changes (as frequency variations) to an external receiver.

Radio pill circuitry

The transmitter

The transmitter converts changes in the transducer variable (e.g. L, C) into frequency variations in the tuned circuit and then radiates these frequency-modulated (f.m.) signals for detection by an external receiver. F.m. signals are more reliable in this application than amplitude-modulated (a.m.) signals, since the latter are severely affected by transmitter location and orientation.

In the typical oscillator circuit, shown in Fig. 8.4(a), the coil L and capacitor C_1 in parallel provide basic oscillations at a natural or resonant frequency. The latter is modulated by a variation in, for example L, which is determined by either pressure or temperature, depending on which of the previous radio pills is being used. The oscillations are maintained by a transistor powered by a small (few volts) battery.

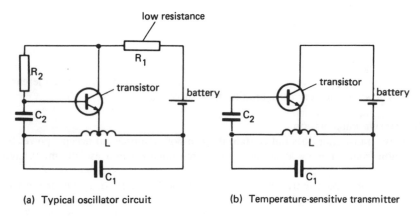

(a) Typical oscillator circuit   (b) Temperature-sensitive transmitter

Figure 8.4   Simple transmitter circuits

Alternatively, the simple circuit of Fig. 8.4(b) may be used to transmit a signal containing both pressure and temperature information. The circuit oscillates at a frequency, dependent on L and C_1, of approximately 1 MHz, and when C_2 becomes charged the transistor is cut off and oscillations cease. C_2 then gradually discharges by leakage through the transistor until the latter again becomes operational and oscillations restart. The oscillations therefore take place in short bursts, a phenomenon referred to as squegging. The frequency of squegging (i.e. the frequency of the bursts) depends on the leakage of the transistor. In germanium transistors this in turn depends on temperature: the higher the temperature, the quicker the oscillations restart. Thus, the final signal consists of periodic bursts of radio-frequency waves in which:

(a) the radio-frequency reflects pressure changes (since L depends on pressure); and

(b) the frequency of the bursts reflects temperature changes.

It is then possible to employ a receiver capable of accepting this compound signal and decoding it into its two separate components with their relevant information. Alternatively, one of the components can be 'blocked' and a simple receiver used to detect the remainder.

The receiver

The receiver's function is to pick up the transmitted f.m. signal and to convert the frequency variations into a d.c. output, which can then be fed to a recorder. Calibration is achieved using a series of known inputs.

If the percentage change in carrier frequency is small and hence difficult to measure accurately, the transmitted signal can be combined with that from a local oscillator of similar but fixed frequency to produce beats (see page 56). The beat frequency (difference in frequency between the two component signal frequencies) reflects changes in the carrier frequency and can be measured accurately.

The power supply

(a) Active transmission

The power supply necessary to maintain oscillations in the transmitter circuit must have a long life (at least a few days) and yield a constant voltage. Single cell batteries provide the simplest solution and those based on zinc and mercuric oxide, known as mercury cells, are the most successful.

(b) Passive transmission

A passive transmitter does not contain its own power source but accepts power by induction from a remote location and then re-radiates it in a usefully modulated form.

In one type of passive transmitter, a grid-dip meter is used to register changes in the frequency of the tuned circuit in the transmitter. An external coil driven by an oscillator of adjustable frequency is placed near the passive transmitter. Some energy is transferred to the passive transmitter by induction and a fall or dip in the external oscillator current is observed. The frequency of the external oscillator is adjusted until the dip in current reaches a maximum, and this corresponds to the passive transmitter resonating with the oscillator frequency. Thus, by cyclically scanning the oscillator frequency, changes in the tuning of the passive transmitter circuit may be followed, and hence the physiological variable in question monitored.

Although passive transmission can suffer from weak and fading signals due to the changing position and orientation of the 'pill', devices using it are very small and light, and have long lifetimes not limited by batteries.

Operating frequency

There are many conflicting factors which influence the choice of basic operating frequency. Smaller components tend to produce higher frequencies, whilst lower frequency units are more stable and the transmissions produced suffer less attenuation.

British Telecom has allocated a special frequency band between 102.2 MHz and 102.4 MHz for biomedical telemetry to reduce interference effects due to local radio stations. These frequencies tend to be used in macrotelemetry, whereas frequencies around 400 kHz are generally selected for short-range work using radio pills.

Uses of biomedical telemetry

Pressures within the blood stream, bladder, uterus, stomach, intestines and head have all been successfully monitored, assisting with diagnoses ranging from ulcers to hydrocephalus. Body core and brain temperatures have been recorded accurately over prolonged periods of time, being of particular importance in extreme-temperature environments.

Advantages of radiotelemetry include the extraction of signals from relatively inaccessible sites, the absence of surface equipment and leads, and the mobility of the 'patient'. Difficulties arise through interference effects, the limited range of the transmitters, and undesirable body reactions (e.g. blood clotting) to the capsule material.

Exercise 8

1 Explain in detail how pressure variations along the gastrointestinal tract can be continuously monitored using a radio pill. How does:
   (a) an increase in pressure in the tract,
   (b) a change in the position and orientation of the pill, and
   (c) an increase in background noise
      affect the final recording?

2 Why is the radio pill sometimes referred to as the 'miniature transformer method of communication'? Why is frequency modulation chosen in preference to amplitude modulation for the transmission of information?
   Give two advantages and two difficulties encountered when using radio pills, and describe two situations in which they are successfully employed.

3 What type of radio pill would be most suitable for recording:
   (a) pressure inside the head in cases of hydrocephalus;
   (b) vaginal temperature fluctuations (associated with ovulation and potentially fertility) over long periods of time;
   (c) blood pressure variations in one of the heart chambers;
   (d) the core temperature of an athlete during training?
   Describe in detail the action of two of these devices.

4 Explain the following terms:
   (a) macrotelemetry;
   (b) endoradiosonde;
   (c) transducer;
   (d) resonant frequency;
   (e) frequency-modulated signal.
   Describe the action of a simple ratio pill which can simultaneously transmit information about temperature and pressure.

9 | Light and electron optics

The electrostatic image intensifier

Luminescence and screening

When ionising radiation (e.g. X-radiation) is absorbed in certain materials the energy raises some of their atoms to excited states. These excited atoms subsequently decay to their ground state re-emitting the energy in the form of visible light photons. Using a microscope, these may be observed as minute flashes of light appearing at random, and they are known as scintillations. The process is known as luminescence and the particular materials in which it occurs are known as scintillators or phosphors. Luminescence falls into two categories, namely:

(a) fluorescence, in which the emission of light occurs within 10^{-8} s of excitation;

(b) phosphorescence, in which the emission of light occurs after a delay of at least 10^{-8} s after excitation.

In X-ray fluoroscopy, X-rays are directed through the body onto a fluorescent screen where the X-ray image of body structures is converted into a visual image. During this so-called screening process, a radiographer can observe body functions such as heart activity over a period of several minutes. However, conventional screens produce dim images and X-ray currents must be strictly limited to reduce radiation risk to the patient. Screening can require the use of a darkened room and time for the radiographer's eyes to become dark-adapted, with their consequent loss of visual acuity (see page 36). In order to improve image brightness without increasing the X-ray current an image intensifier may be used, making dark surroundings unnecessary.

The electrostatic image intensifier

The X-ray pattern from the patient is directed onto the input or pick-up fluorescent screen of the image intensifier (Fig. 9.1) where a visual image is produced. In close optical contact with the screen is a transparent surface (e.g. of caesium–antimony) known as the photocathode. Light from the fluorescent screen stimulates the photocathode to emit photoelectrons in a pattern duplicating that of the original X-rays.

Using an electron lens system to focus the electron beam and a final accelerating electrode (held at $\approx +25\,\text{kV}$ with respect to the input screen) to accelerate the electrons, the beam is directed towards a small fluorescent screen known as the output or viewing screen at the other end of the evacuated tube. Here, an inverted

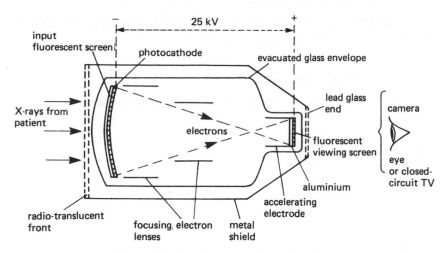

Figure 9.1   The electrostatic image intensifier

reduced visual image of greatly increased brightness is produced and may be observed through a lead glass window in the outer metal shield. A thin layer of aluminium on the photocathode side of the viewing screen permits the passage of electrons, but stops any light produced at the viewing screen from passing back to the photocathode.

Zinc–cadmium sulphide crystals evenly spread across and attached to a cardboard surface forms a typical input screen, yielding a bright image with minimum 'afterglow' which ensures a sharp image. When the output screen is to be observed directly, a popular phosphor here is silver-activated zinc sulphide, with its green-yellow emission matching the maximum colour sensitivity of the eye, and its short but noticeable afterglow combining well with the eye's persistence of vision. A shorter afterglow is desirable, on the other hand, if the viewing screen is linked to a camera or closed-circuit television, and other phosphors with blue–white emissions are then more common.

Intensification of the image

The overall intensification of the image is due to:

(a)   reduction in size of the image: a reduction in linear dimensions of about 5:1 is typical, and results in an increase in intensity of 5^2 $(=25)$;

(b)   increase in energy of the electrons as they are accelerated, and their consequent conversion into more light energy (greater brightness) at the viewing screen. This may result in intensity increases exceeding 100.

The total intensification of the image in a typical image intensifier may be of the order of 2500 (i.e. 25×100) or even higher.

Since each point on the input image is 'intensified' by an equal amount, the contrast and resolution are not strictly speaking improved by the instrument itself; that is, the image intensifier cannot produce detail not already present in the input image. However, since the eye's resolution is better at higher light intensities, the final resolution achievable is in fact improved.

Closed-circuit television

The basic components of the system (see Fig. 9.2) perform briefly as described below.

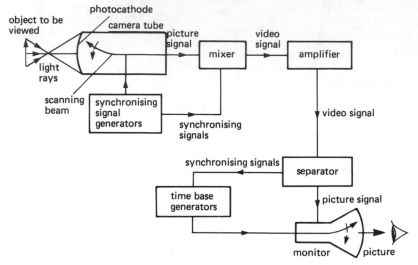

Figure 9.2   A closed-circuit television system

Camera tube

The object to be viewed is first focused onto the photocathode of a television camera tube and a pattern of electrostatic charges, corresponding to the original optical image, is built up over the surface of the photocathode. The latter is then scanned by an electron beam which gives rise to a fluctuating output current, called the picture signal.

Synchronising signal generators

The vertical and horizontal deflections of the scanning beam are controlled by two time-base generators and the frequency at which they operate is determined by pulses from the synchronising signal generators.

Mixer

It is essential that the scanning process in both camera tube and monitor are kept synchronised with each other. The synchronising signals are therefore super-imposed on the picture signal in the signal-mixing circuit.

Amplifier

The composite signal, called the video signal, is then amplified and fed through a cable to the cathode ray tube. There is considerable flexibility in this part of the system. The cable can be long, linking different rooms or even buildings, and if required the same video signal can be fed simultaneously to a number of monitors.

Separator
This separates the picture signal from the synchronising signals. The picture signal is then fed to the monitor (cathode ray tube).

Time-base generators
The synchronising signals are fed into the time-base generators which control the scanning beam in the monitor. These signals thus hold the scanning systems of the camera tube and monitor in synchronisation, so that both beams scan corresponding elements of their respective images.

Monitor
This is a conventional cathode ray tube, consisting very simply of an indirectly heated cathode yielding electrons by thermionic emission, and a system of electron lenses to focus the electron beam onto a luminescent screen. The electron beam may be deflected using electrostatic deflector plates, or using magnetic deflecting lenses, the latter being almost universally adopted in television cathode ray tubes. The particular viewing screen used depends on the method of observation (see page 107) but is typically yellow-green emissive with short afterglow.

The picture signal energises the deflecting lenses, thus resulting in a final screen image duplicating the original optical image. Each monitor has its own controls for the adjustment of image contrast and brightness.

In an X-ray closed-circuit television system, the original object viewed is the output screen of an image intensifier. The system offers several advantages over the use of the intensifier alone. It is more sensitive, produces a screen image of greater brightness and resolution at lower exposure rates, and can provide contrast enhancement. In addition to its use in screening (see page 106), the system can be invaluable during surgery to assist, for example, in the positioning of radio-opaque catheters, needles, and so on.

Other applications of closed-circuit television include the remote observation of patients and equipment in intensive care units and radiotherapy treatment rooms, continuous recording using video-tape, and the use of monitors, both large and small, in consultation and teaching.

Fibre optics

Fibre-optic bundles
If light enters the end of a solid glass rod so that the light transmitted into the rod strikes the side of the rod at an angle θ, exceeding the critical angle, then total internal reflection occurs (see Fig. 9.3). The light continues to be internally reflected back and forth in its passage along the rod, and it emerges from the other end with very little loss of intensity.

This is the principle of fibre optics, in which long glass fibres of very small cross-sectional area transmit light from end to end, even when bent, without much loss

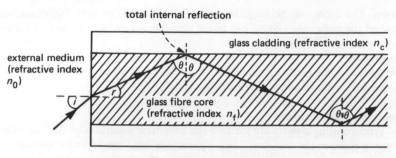

Figure 9.3   Glass fibre with cladding

of light through their side walls. Such fibres can then be combined into 'bundles' of dozens to thousands of fibres for the efficient conveyance of light from one (often inaccessible) point to another.

If the glass fibre comes into contact with a substance of equal or higher refractive index, such as an adjacent glass fibre, dirt or grease, then total internal reflection does not occur and light is lost rapidly by transmission through the area of contact. To avoid such 'leakage' and to protect the fibres, they are clad in 'glass skins' of refractive index lower than that of the fibre core.

Using the notation in Fig. 9.3, as the angle of incidence i increases, r increases and θ $(=(\pi/2)-r)$ decreases. Eventually, θ reaches c, the critical angle, and any further reduction in θ results in transmission through the side wall. The limiting case, for which total internal reflection *just* occurs, corresponds therefore to a maximum value of i known as i_{max}. Using:

$$n \sin i = \text{constant}$$
$$n_0 \sin i_{max} = n_f \sin r$$
$$= n_f \sin\left(\frac{\pi}{2} - c\right)$$
$$= n_f \cos c$$
$$= n_f \sqrt{(1 - \sin^2 c)} \qquad [9.1]$$

Also:

$$n_f \sin c = n_c \sin 90°$$
$$\therefore \qquad \sin c = \frac{n_c}{n_f}$$

Substituting this into equation [9.1] gives:

$$n_0 \sin i_{max} = n_f \sqrt{\left(1 - \frac{n_c^2}{n_f^2}\right)}$$
$$n_0 \sin i_{max} = \sqrt{(n_f^2 - n_c^2)}$$

The expression $n_0 \sin i_{max}$ is called the numerical aperture of the fibre. A typical value for this might be 0.55, making i_{max} about 33° in air. Sometimes i_{max} is referred to as the half-angle of the fibre, since it describes half the field of view acceptably transmitted. The numerical aperture (and hence i_{max}) can be increased by using a core of high refractive index. However, these glasses have a lower efficiency of transmission, especially at the blue end of the spectrum, and are not commonly used.

The above analysis applies only to a straight fibre. If the fibre is curved, the angles of incidence vary as the light travels along the fibre and losses occur if the angles fall below the critical angle. In practice, a radius of curvature down to about twenty times the fibre diameter can be tolerated without significant losses.

Coherent and incoherent bundles

An ideal fibre transmits light independently of its neighbours, so if a bundle of fibres is placed together in an orderly manner along its length, with the relative positions remaining unchanged, actual images may be transmitted along the fibre (Fig. 9.4). Such an arrangement is called a coherent bundle, and consists of fibres of very small diameter, about 10 μm. The ends of the bundle are cut square and polished smooth to prevent distortions. Each fibre transmits a small element of the image which is seen at the other end of the coherent bundle as a mosaic. The eye has to 'look through' the fragmented structure to appreciate a clear image.

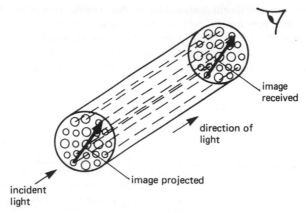

image received

direction of light

image projected

incident light

Figure 9.4   Coherent fibre-optic bundle

The image to be transmitted is either in direct contact with the end of the bundle or focused on to this surface. The image formed at the other end is viewed using an eyepiece incorporating magnification. One novel method of magnification is to make one end of the fibres smaller than the other. For example, if they have an average diameter of 5 μm at the image end, and 50 μm at the viewing end, a magnification of $\times 10$ is achieved.

In contrast, a bundle of fibres arranged at random is known as an incoherent bundle (or sometimes simply a light guide) and is suitable only for the transport of light not of images. The fibres of such a bundle are relatively large, having diameters of about 50–100 μm.

Transmission efficiency and resolution

A light beam inevitably suffers some attenuation in its passage along a fibre core: a 50 per cent loss for every 2 m travelled is typical. In addition, there are 'end losses', which are light losses at the end faces due to partial reflection and incidence on the cladding material. Thus, light sources need to be very powerful, and even then problems can arise when viewing coloured images since different wavelengths have different transmission efficiencies.

The thinner and more numerous the fibres, the greater should be the resolution. However, when the core diameter falls below about 5 μm diffraction starts to occur and transmission efficiency drops. Hence, although fibres with core diameters down to about 1 μm have been used, typical diameters are nearer 10 μm.

A deterioration in image quality may occur for a number of reasons, for example defects in the end faces of the fibres, misalignment of fibres, broken fibres (causing black spots), or light leakage between adjacent fibres (producing 'cross-talk').

Fibre-optic endoscopy

Introduction

An endoscope is an instrument designed to provide a direct view of an internal part of the body, and possibly to perform tasks such as the removal of samples, injection of fluids and diathermy. Fibre optics has extended the scope of the instrument considerably by permitting the transmission of images from inaccessible areas such as the oesophagus, stomach, intestines, heart and lungs.

The fibre-optic endoscope

The long flexible shaft of the instrument (see Fig. 9.5) is usually constructed of steel mesh, often with a crush-resistant covering of a bronze or steel spiral. It is then sheathed with a protective, low-friction covering of PVC or some other impervious material, which forms a hermetic seal around the instrument. The shaft is about 10 mm in diameter, about 0.6–1.8 m long (depending on the application) and has a short deflectable section about 50–85 mm long leading to its distal tip (Fig. 9.5 inset).

Within the shaft lie:

(a) at least one non-coherent fibre-optic bundle to transmit light from the distant light source to the distal tip;
(b) a coherent fibre-optic bundle transmitting the image from the objective lens at the distal tip;
(c) an irrigation channel through which water can be pumped to wash the objective lens;
(d) an operations channel for the performance of tasks;
(e) control cables.

The viewing end of the endoscope contains:

(f) an eyepiece, with focus controls and camera attachment;

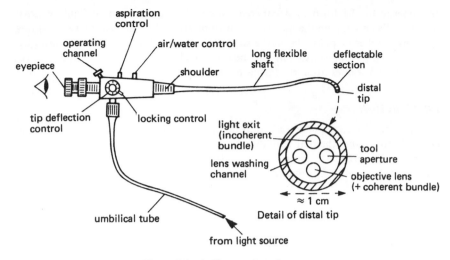

Figure 9.5   A fibre-optic endoscope

(g) distal tip deflection controls, giving polydirectional control up to about 200°, plus locking capability;

(h) objective lens control which may be a push–pull wire effecting focusing;

(i) valve controls for air aspiration, (suctioned withdrawal of body fluids through the operations channel) and lens washing or air insufflation (application of water or air jet through the irrigation channel);

(j) operating channel valve, which controls the entry of catheters, electrodes, biopsy forceps and other flexible devices;

(k) connection with the umbilical tube, providing light through a non-coherent fibre-optic bundle and water or air from the pump or aspirator system.

Some applications of fibre-optic endoscopy

Endoscopic examination of the gastrointestinal tract has proved especially successful with the diagnosis and treatment of ulcers, cancers, constrictions, bleeding sites, and so on. The heart, respiratory system, and pancreas have also been investigated.

Another application is the measurement of the proportion of haemoglobin in the blood which is combined with oxygen using an oximeter. Two incoherent bundles are introduced into the blood stream: one is used to illuminate a sample of blood and the other to assess the absorption of light by the blood.

Lasers in medicine

The laser

The word laser is an acronym derived from Light Amplification by Stimulated Emission of Radiation. The instrument operates basically by absorbing non-coherent radiation from, for example, a xenon arc lamp and re-emitting some of this energy in the form of a coherent, monochromatic beam of light of great intensity. Depending on the material used to produce the laser action, the resulting beam may either be pulsed (as in the solid-state ruby laser) or continuous (as in the helium–neon gas laser). A continous beam provides much smaller output powers than the peak values attained during pulse operation.

Laser action

When in an electromagnetic radiation field, atoms of a material may absorb and emit photons, and thermal equilibrium is reached when as many photons per second are absorbed as emitted. In this situation, there are far more atoms in the ground state (energy E_0) than in excited states (energies E_1, E_2, and so on) and such a distribution is called a normal population (see Fig. 9.6(a)).

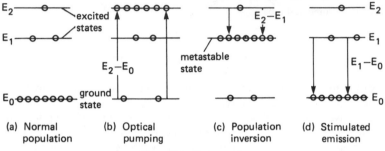

Figure 9.6   Laser action

In certain materials, it is possible to get an inversion of this population using a technique known as optical pumping (Fig. 9.6(b)). The atoms are first optically excited, for example by using light focused from a xenon arc lamp, to the energy state E_2. Almost immediately, they 'decay' or re-emit energy and reach the state E_1. The latter is a metastable state (i.e. decay from such a state is more delayed) and so an accumulation of atoms in this state occurs giving a population inversion (Fig. 9.6(c)). This is just what is required for laser action. If a photon of energy $(E_1 - E_0)$ is now introduced to the system, the excited atoms are stimulated to return to the ground state by emitting photons of precisely this energy $(E_1 - E_0)$, (Fig. 9.6(d)). Since all the emitted photons are in phase, the resulting beam of light is not only monochromatic but also very coherent and extremely intense.

Amplification may be increased by using mirrors (or silvered surfaces of the crystal in the solid-state laser) at the ends of the material. The photons thus suffer multiple reflections within the system stimulating the emission of more and more photons during their passage. A very powerful and narrow beam of light eventually emerges through a section at one end which is only partially silvered.

Medical applications of lasers

(a) Treatment of tumours
Malignant tissue absorbs laser radiation more strongly than does healthy tissue. Hence, tumours may be treated without causing undue damage to the surrounding

normal tissue. Laser beam properties (for example, wavelength, power density, energy dose and number of pulses) are chosen to suit the absorption properties of the particular tumour and varying success has been achieved. Pigmented or dark tumours which absorb most light are readily destroyed; red-reflecting vascular tumours are little affected by the red radiation from a ruby laser; white tumours are not successfully treated.

Superficial tumours can be treated directly. The removal of tattoos and the lightening of 'port-wine' type birth marks is also possible. For deeper tumours, the area has to be exposed surgically or irradiated through a fibre-optic endoscope (see page 112).

(b) Retinal photocoagulation

If a tear develops in the retina, fluid can pass from the vitreous body through the hole pushing the retinal cells away from the choroid and causing partial blindness. In a process known as retinal photocoagulation, the retina may be 'spot-welded' back in position using a short pulse from a ruby laser. A typical pulse would deliver an energy of about 0.1 J in less than 1 ms over an area of about 10^{-3} mm^2. Hence, very few retinal cells are damaged and pain is not experienced since the threshold for the sensation of pain is about 200 ms. Normal vision is restored as long as treatment is prompt.

(c) Surgical applications

A laser beam can cut through tissue and blood vessels without causing bleeding: the beam cauterises blood vessels as fast as it severs them. Not only does this make surgery easier, but it also permits operations like the removal of a portion of a liver lobe which were previously impossible due to the inevitable haemorrhage.

(d) Dental studies

Healthy teeth are normally smooth, white, highly reflective and hence poor absorbers of laser radiation. In contrast, dental caries (decayed areas) are dark, dull, have rough surfaces and consequently are good absorbers of laser radiation. Therefore, if a laser beam is applied to teeth, the caries selectively absorb the radiation and are 'drilled out' as necessary. A ruby laser delivering about $1\,\mathrm{Jm}^{-2}$ at the caries is suitable.

(e) Other applications

The laser has been used as a lining-up device in X-ray radiotherapy; as a vaporising tool in the spectrographic analysis of samples from teeth, bone, internal organs, and so on; and in conjunction with a fibre-optic endoscope, its power to vaporise obstructions like blood clots and stones holds great promise for future applications.

Exercise 9

1 Using a clear labelled diagram, describe the principles of operation of an image intensifier. What factors affect (a) the brightness and (b) the sharpness of the final image?

2 Describe the essential features of a closed-circuit television system.

In a given fluoroscopic examination, the fluorescent screen may be linked either to an image intensifier or to a closed-circuit television system. Discuss the advantages of using each system.

3 How is the principle of total internal reflection used in fibre optics? What conditions must be met to ensure the efficient transmission of light along a fibre-optic bundle?

(a) A glass fibre core of refractive index 1.5 is clad in a material of refractive index 1.4. Light falls on the end of the core at an angle of incidence of 20°, and is refracted into the core. If the fibre is straight, calculate the angle of incidence at which the light internally strikes the side of the core, and hence show that total internal reflection will occur.

(b) The fibre is now bent so that the internal angle of incidence on the side decreases. Calculate the maximum angle of bending before total internal reflection ceases to occur. (You may assume the bending occurs in one abrupt process. Discuss how it would occur in practice and how this would affect your answer).

4 Explain the following terms used in fibre optics:

(a) critical angle;

(b) numerical aperture of a fibre;

(c) coherent bundle of fibres.

A particular fibre has a refractive index of 1.52 and is clad in a material of refractive index 1.43. Find:

(d) the numerical aperture of the fibre;

(e) the half-angle of the fibre in air.

5 Explain carefully the factors which affect:

(a) the brightness,

(b) the detail,

(c) the quality,

of the image viewed using a coherent fibre-optic bundle.

Such a coherent bundle consisting of fibres of core diameter 12 μm, each protected by a cladding of thickness 2 μm, is uniformly illuminated at one end. Assuming that absorption in the core material leads to a reduction in light intensity of 20 per cent for every 0.5 m travelled, estimate the percentage of the incident light intensity reaching the end of a fibre-bundle 1.5 m long. (You may neglect any partial reflection of the incident light beam at the end of the bundle.)

6 Describe the basic structure of a fibre-optic endoscope.

Discuss how the use of ancillary equipment extends its mode of operation from one of simple observation.

7 Discuss the energy changes which take place in the production of a beam of laser radiation.

Describe three clinical applications of such a beam.

A laser beam is focused on to a spot of diameter 2 mm. If the beam can deliver energies ranging from 0.5–360 J in bursts lasting from 0.5–3 ms, find the possible range of power density (power delivered per unit area).

10 Ultrasonics

Properties of ultrasound

Ultrasonic sound waves, or ultrasound, have frequencies above the human ear's audible range, that is, greater than 20 kHz. Their properties are those common to all sound waves (see page 47).

Of particular interest in the medical field is the reflection of ultrasound. In diagnostic studies, a beam of ultrasonic waves (in the frequency range 1–15 MHz) is directed into the body and the reflections or echoes from different body interfaces are detected and analysed, yielding information about these structures. Important parameters in such studies are:

(a) intensity reflection coefficient $\alpha_r = \left(\dfrac{Z_2 - Z_1}{Z_2 + Z_1}\right)^2$ (See [4.3] on page 48)

    where $Z (= \rho c)$ is specific acoustic impedance;

(b) absorption coefficient k, where $I_x = I_0 e^{-kx}$ (See [4.2] on page 47)

Average values of Z and k for some common biological materials are given in Table 10.1.

Table 10.1 Ultrasound in biological materials

Medium	Density (ρ) (kg m^{-3})	Ultrasound velocity (c) (m s^{-1})	Specific acoustic impedance $Z (= \rho c)$ (kg m^{-2} s^{-1})	Absorption coefficient k at 1 MHz (Bm^{-1})
Air	1.3	330	429	120
Water	1000	1430	1.43×10^6	0.02
Blood	1060	1570	1.59×10^6	2
Brain	1025	1540	1.58×10^6	9
Fat	952	1450	1.38×10^6	6
Muscle (average)	1075	1590	1.70×10^6	23
Bone $\begin{cases} \\ \end{cases}$ (varies)	1400 1908	4080	$\begin{cases} 5.6 \times 10^6 \\ 7.78 \times 10^6 \end{cases}$	130

Some typical interfaces encountered in the body are:

(a) two soft tissues; the difference in Z is small and only about 1 per cent of the incident intensity is reflected back ($\alpha_r \approx 0.01$) but this is sufficient to produce a detectable echo;

(b) bone–soft tissue; the difference in Z is larger, leading to more intense echoes;

(c) air–soft tissue; the difference in Z is enormous, and most of the incident intensity is reflected back. This not only makes transmission of the ultrasound from the transmitter into the body difficult, but also prevents examination of structures beyond the lungs because the air present effectively 'blocks' further transmission.

Ultrasonic absorption in the body depends on many factors including frequency, temperature, and density of the medium. An average attenuation is about $10 \, \text{Bm}^{-1} \, \text{MHz}^{-1}$, indicating that higher frequencies are less useful for deep penetration studies.

Piezoelectricity

The piezoelectric effect describes an interchange between mechanical and electrical energy which occurs in certain crystals, known as piezoelectric crystals such as quartz or the synthetic ceramic, lead zirconate titanate. In an unstressed state, the centres of symmetry of both the positive and negative ions of such a crystal lattice coincide and no effective charge appears on electrodes attached to the crystal (see Fig. 10.1(a)). However, when the crystal is compressed or extended (Fig. 10.1(b) and (c)), the centres of symmetry move, no longer coincide, and give rise to free charges on the electrodes thus producing a voltage across them.

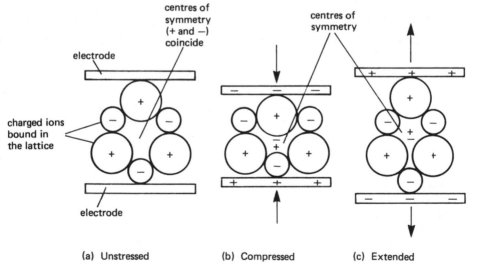

Figure 10.1   The piezoelectric effect

Conversely, if a voltage is applied across an unstressed piezoelectric crystal, the centres of symmetry move, hence deforming the crystal. An applied alternating voltage thus gives rise to mechanical vibrations in the crystal, a maximum response (or resonance) occurring when the applied frequency matches a natural frequency of vibration of the crystal.

A piezoelectric transducer can operate as:

(a) an ultrasound generator, by applying a stimulating voltage of suitable frequency, thereby causing crystal vibrations and the emission of ultrasonic waves,
(b) an ultrasound detector, by monitoring the piezoelectric voltage developed across the crystal when it is forced to vibrate by incoming ultrasonic waves.

A typical transducer used in medical applications is shown in Fig. 10.2. The electrodes must be light and they usually consist of thin layers of silver. The one nearest the patient is connected to the earthed metal case for safety. The other is called the 'active' or 'live' electrode and is connected via a coaxial cable to either a source of power (when used as a transmitter) or an amplifier and cathode ray tube (when used as a receiver).

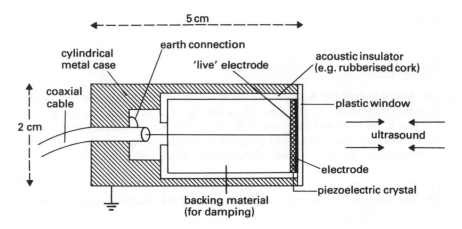

Figure 10.2   A piezoelectric transducer

Clear distinct echoes are obtained using short pulses. To generate such pulses, the vibrations must be damped out as quickly as possible. This is achieved by bonding a damping material (araldite or other epoxy resin) to the back face of the crystal.

Ultrasonic scanning

Introduction

Diagnostic information about internal structures may be obtained by transmitting short sharp pulses of ultrasound from a transducer into the body and then detecting and analysing the pulses which are reflected back.

In a typical pulse–echo diagnostic procedure, the maximum mean ultrasound power delivered is about 10^{-4} W, and the frequency is in the range 1–5 MHz.

Coupling medium

A coupling medium is essential between the transducer and body surface to prevent the excessive reflections which occur at any air—solid boundary (see page 117).
Early techniques involved the immersion of patient and transducer probe in a
water bath. Later equipment used a probe in a plastic bag in close contact with the
skin. The bag was filled with water, sometimes with the addition of chemicals, to
provide a good acoustic match with the skin. Disadvantages of this method
include the extra pressure on the patient, the acoustic loss in the water and
reflective losses at interfaces, particularly if any air is trapped between the plastic
bag and the patient. Its major advantage, however, is that any movement of the
probe (to scan the body) causes no local disturbance of the body surface with
which it does not come into contact.

The coupling medium most commonly used today is a film of oil, smeared on
the patient's skin. This is quick and simple to apply and involves less acoustic loss.
Alternatively, a water-based cellulose jelly is used, particularly in examinations of
the head where air can get trapped near the hair roots.

The A-scan

The A-scan system (or A-scope, Fig. 10.3) is basically a range-measuring system.
It operates by recording the time taken for an ultrasonic pulse to travel to an
interface in the body and then be reflected back. The time-measuring instrument is
the cathode ray tube which must therefore be synchronised with the transmitter/
receiver system. The components of the system are described below.

(a) The rate generator (synchroniser)
This simultaneously triggers (a) the transmitter; (b) the time-base generator; (c) the
swept gain generator.

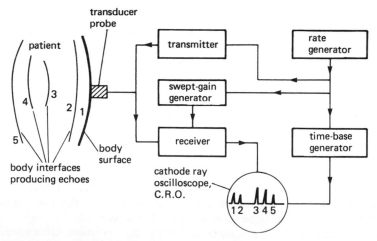

Figure 10.3 The A-scan system

(b) The transmitter
This energises the piezoelectric transducer which then emits a short (few microseconds) sharp pulse of ultrasound into the body.

(c) The time-base generator
This supplies a voltage across the X-plates (horizontal deflection) of the CRO so that the spot starts moving steadily across the screen.

(d) Receiver
Voltages which appear at the transducer probe, either due to the transmitted pulse or reflected pulses returning from the patient, are amplified at the receiver and applied to the Y-plates (vertical deflection) of the CRO. A vertical line thus appears on the screen each time an echo returns to the probe. The positions of these lines on the screen (e.g. 1, 2, 3, 4, 5) correspond to the positions of the associated reflecting surfaces in the body (i.e. 1 (body surface), 2, 3, 4, 5).

(e) Swept-gain generator
Due to attenuation of the ultrasound pulses in the body, echoes from deeper interfaces tend to be very weak. Hence, these are amplified more than those originating close to the probe using the swept-gain generator. This increases the gain of the receiver with time at an appropriate rate.

(f) Cathode ray oscilloscope
One complete movement of the spot across the screen corresponds to the transmission of one pulse and the receipt of a number of echoes. If this entire process is repeated very rapidly a continuous trace is seen on the screen. The CRO may then be used as a time-measurer, allowing estimates of the time elapsing for example between positions 1 and 2 on the screen. This corresponds to the time taken for an ultrasound pulse to travel from the probe to the reflecting surface 2 and back again. Then, knowing the velocity of the pulse ($\approx 1500 \, \text{m s}^{-1}$) the depth of the interface 2 below the body surface may be evaluated. The CRO may be calibrated to give distances between reflecting interfaces in the body directly.

The B-scan (or B-scope)

In the B-scan, the echo signals are not applied to the Y-plates of the CRO but instead are used to control the brightness of the spot on the screen. Hence, the static B-scan displays the range of reflecting surfaces using spots whose brightness give a measure of echo amplitude.

By scanning the patient, a number of static B-scans can be combined into a two-dimensional compound B-scan (Fig. 10.4). The ultrasonic probe is mounted on a mechanical scanner, which allows movement in two dimensions under manual control. Voltages from potentiometers record the position and orientation of the probe. These readings are then fed to the X- and Y-plates of the CRO and synchronise the position of the spot with the corresponding point of the body being examined. As the probe is moved around the patient, the display is continuously recorded either using a storage CRO (see page 122) or photo-

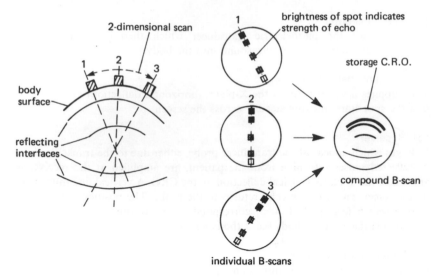

Figure 10.4   The compound B-scan

graphically using Polaroid film, and hence a cross-sectional picture in the plane of the scan (a tomograph) is constructed in two dimensions.

The probe is usually oscillated, or rocked, as it is moved around the patient. This increases the probability that the pulse will strike a typically irregular interface at normal incidence and so produce a strong echo. This rocking of the probe is a skilful technique during which the probe must be kept in intimate contact with the oil-covered skin, to prevent the introduction of any air space with its consequent excessive reflections.

Time-position scanning

A static B-scan may be modified by applying a low-velocity time-base generator across e.g. the Y-plates of the CRO. During the sweep time ($\approx 3\,s$) the B-scan is moved vertically at a constant low velocity. Thus, if any of the examined reflecting interfaces move during this time, their movement creates horizontal deflections in the vertical line pattern obtained (Fig. 10.5). In particular, a pulsating structure such as the heart gives rise to a regular pulse pattern on the screen. Once again, the recording is made using either Polaroid film, or a storage CRO.

The storage CRO (electronic storage tube)

The essential features of a cathode ray oscilloscope capable of storing information, for example during a scan or simply for subsequent study, are shown in Fig. 10.6. A writing gun deposits the charge pattern corresponding to the signal investigated on the storage surface, a metal mesh backing electrode coated with an insulator.

To display the stored pattern, the flood gun emits low-velocity electrons towards the screen. These electrons penetrate the storage mesh and are accelerated towards

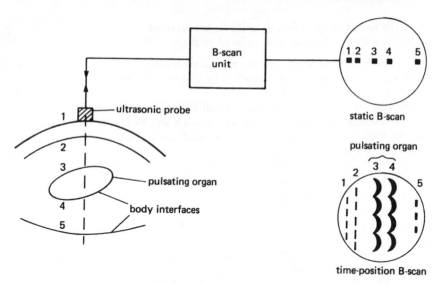

Figure 10.5 Time–position scanning

the screen to form an image. The brightness of the image is determined by the stored charge pattern which is thus visually reproduced.

The image is erased by the application of a small positive potential to the backing electrode in the storage surface.

Gray-scale systems

For pulses of a given power emitted by the transducer, the major factors which affect the intensity of the received ultrasound signals are:

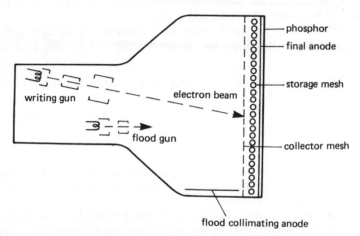

Figure 10.6 Storage cathode ray oscilloscope

(a) the nature of the reflecting surfaces encountered;
(b) the amount of attenuation suffered en route.

The latter depends largely on ultrasound path length in the body, deeper sites returning weaker echoes. Compensation for this may be achieved using the swept-gain generator, (see page 121).

The reflecting interface properties which determine echo strength include the change in specific acoustic impedance (Z) across it and its regularity. Strong echoes are obtained by striking perpendicularly interfaces between media with large Z differences (e.g. bone and soft tissue). In a typical clinical scan a range of about 90 dB (a factor of 10^9) exists between the size of the smallest and largest echoes! Before the advent of the gray-scale system of imaging in 1974, this resulted in strong echoes all being displayed at similarly high intensities whilst valuable small echoes failed to register at all.

The gray-scale system employs selective amplification of low level echoes, which originate from within soft tissues, and displays these echoes at the expense of larger echoes. Subtle soft tissue textural differences, previously lost, are now included to give a more informative final image.

Resolution

The resolution (or fineness of detail) achievable in the ultrasound image depends very much on the axis considered. The axial plane is a plane containing the ultrasound beam and resolution here is determined by the brevity of the ultrasound pulse. Heavy damping in the transducer (see page 119) produces pulses of about 1 µs duration, with a resulting axial resolution of approximately 1 mm. This resolution suffers if longer pulses are used.

Lateral resolution (in a direction perpendicular to the beam) depends largely on beam width and hence varies with distance from the transducer. The frequency of the beam, its spatial and directional characteristics, as well as the particular imaging technique used, also influence lateral resolution. For transducers in common clinical practice, lateral resolutions of 3 mm are typical, although resolutions approaching 0.1 mm are achievable.

In addition to these limitations set by the instrumentation, all scans suffer further blurring or confusion due to spurious signals or *artifacts*. These arise for a number of reasons:

(a) When a reflected pulse arrives back at the transducer most of the energy is transmitted to the receiver for subsequent registration on the display. However, some of the energy pulse is reflected back into the body for another 'round trip'. When it next appears back at the transducer an artifact is displayed on the screen. Such *multiple reflection artifacts* are usually recognised by the regularity of their spacings.

(b) The body does not consist of smooth, reflecting interfaces and homogeneous media. Artifacts due to irregularities or, more commonly, the presence of gas can confuse the resulting scan.

(c) Except when internal movement is being studied, as in time–position scanning, any movement of the patient can blur the scan. Such movements include the involuntary movements of vital organs, movement of the foetus in the uterus, and so on.

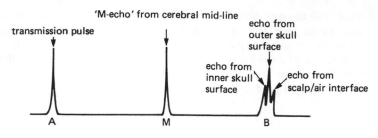

Figure 10.7 Typical echoencephalogram

Applications of ultrasonic scanning

Echoencephalography

Ultrasonic scanning of the brain may lead to the diagnosis of numerous disorders. An A-scan is employed, a typical result being illustrated in Fig. 10.7. The characteristic M-echo is easily distinguished because of its high amplitude which is due to reflection from several co-planar interfaces at about the same distance from the probe. Normally, M should lie half-way along AB, and any departure from this indicates some abnormality, for example, bleeding into one side of the brain due to injury or stroke, intracranial tumours, hydrocephalus.

Obstetrics

The bi-parietal diameter is the maximum diameter of the head, measured between the two parietal bones (at the side of the head above the ears). The measurement of this diameter in the foetal head may be used to study its progressive growth, providing an estimate of the length of gestation and predicting a delivery date, and indicating any possible abnormalities, such as hydrocephalus.

A B-scan is first used to identify the appropriate section through the foetal skull. The ultrasound beam is then carefully directed to pass through the widest diameter of the head, striking the mid-line normally. The distance between the two parietal (skull) bones is then measured accurately on an A-scan.

Ultrasound is successfully used to locate the placenta, the organ on the wall of the uterus through which the foetus is nourished. If the placenta is across the exit from the uterus (placenta praevia) a normal delivery is difficult and a Caesarian section is recommended. Other difficulties include separation of the placenta from the uterine wall endangering the foetus, and a thick, opaque placenta possibly indicating a rhesus baby.

Being harmless to mother and baby, ultrasound finds many applications in obstetrics. Multiple pregnancy, ectopic pregnancy (growth of the foetus outside the uterus) and breech presentations are amongst the conditions easily diagnosed.

Ultrasound cardiography (UCG)

The front of the heart is largely covered by the lungs and pleura, which effectively 'block' ultrasound due to the presence of air. However, there is a small 'window' in the thoracic wall through which the heart may be investigated using ultrasound. The main system used is time–position scanning. Considerable experience is necessary to register and interpret a good trace from a heart valve, since the valve is by no means a plane reflector and the echo amplitude is very dependent on angulation. In addition, it is important to relax the subject first, since fast heart rates are less likely to produce a successful ultrasound cardiogram (UCG).

Doppler methods

The Doppler effect

Consider a source of sound of frequency f in a medium in which the wave velocity is c. In 1 s, f waves are emitted and occupy a distance c (Fig. 10.8 (a)). If the source

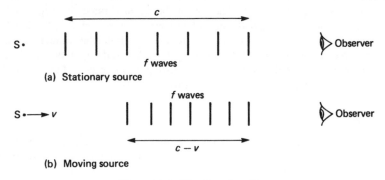

Figure 10.8   The Doppler effect

now moves towards a stationary observer with a velocity v, the f waves emitted in 1s only occupy a distance $(c-v)$ (Fig. 10.8 (b)).

As far as the observer is concerned the initial wavelength of the sound when the source is stationary is given by:

$$\lambda = \frac{c}{f}$$

The apparent wavelength, when the source is moving, is given by:

$$\lambda' = \frac{c-v}{f}$$

Hence, the observer registers an apparent frequency:

$$f' = \frac{c}{\lambda'} = \frac{cf}{c-v}$$

which is higher than the true frequency f. The apparent change in frequency, known as the Doppler shift is given by:

$$\Delta f_1 = f' - f$$

$$= \frac{cf}{c-v} - f$$

$$\therefore \quad \Delta f_1 = \frac{fv}{c-v}$$

When $v \ll c$, then:

$$\Delta f_1 \approx \frac{fv}{c} \qquad\qquad [10.1]$$

Consider now an observer moving at a velocity v towards a stationary source. The f waves emitted by the source in 1s still occupy a distance c, and their apparent wavelength to the observer is still $\lambda = c/f$. However, the velocity of the waves relative to the observer is now $c + v$, so that the apparent frequency becomes:

$$f' = \frac{c+v}{\lambda} = \frac{(c+v)f}{c}$$

The Doppler shift in this case is then:

$$\Delta f_2 = f' - f = \left(\frac{c+v}{c}\right)f - f$$

$$\therefore \quad \Delta f_2 = \frac{fv}{c} \qquad\qquad [10.2]$$

Ultrasound reflection from moving surfaces

Consider an ultrasonic transducer transmitting waves of frequency f towards a reflecting surface, which is moving towards the transducer with a velocity v. The transducer is then further employed to receive the waves reflected from the surface (see Fig. 10.9).

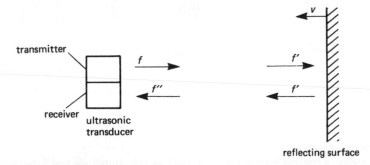

Figure 10.9   Ultrasound reflection from a moving source

The total Doppler shift which occurs is the sum of two parts:

(a) From the standpoint of the reflecting surface, the apparent frequency of the waves reaching it is f', resulting in a Doppler shift, given by equation [10.2], of:

$$\Delta f_2 = f' - f = \frac{fv}{c}$$

(b) The reflection of ultrasound from the surface is equivalent to the transmission of a frequency f' from a virtual source moving with the reflecting surface. Thus, to an observer fixed at the receiving crystal, the apparent frequency is f'' and the additional Doppler shift is, from equation [10.1]:

$$f'' - f' = \Delta f_1 = \frac{fv}{c}$$

assuming $v \ll c$, which is generally true for body reflectors. The total Doppler shift occurring is hence:

$$f'' - f = \Delta f_1 + \Delta f_2$$

$$\therefore \qquad \Delta f = \frac{2fv}{c} \qquad\qquad [10.3]$$

If an ultrasound transducer, or probe, is placed on the body, any moving surface in the body acting as a reflector causes a Doppler shift, Δf, which may be measured and so provide an estimate of v, the velocity of the moving surface relative to the body surface (and hence the transducer).

This is the principle of the continuous wave Doppler system the essential components of which are illustrated in Fig. 10.10 and described below.

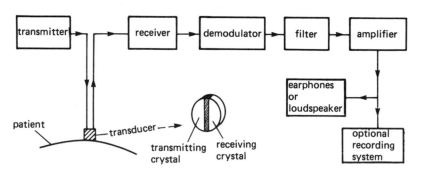

Figure 10.10 Doppler investigation of moving surfaces in the body

(a) The transmitter
This provides a continuous voltage output of constant amplitude and frequency, the latter generally being about 2 MHz.

(b) The transducer

This contains separate transmitting and receiving piezoelectric crystals. If only one crystal was used, signals travelling straight from the transmitter–receiver crystal to the sensitive receiver amplifier might overload the latter.

(c) The receiver

The output contains signals of:

(i) the same frequency as the transmitter (due to stationary reflectors);
(ii) the Doppler-shifted frequency (due to the moving reflector).

(d) The demodulator

This 'mixes' the frequencies to produce beats. The frequency of the beats is equal to the difference between the two input frequencies, which in turn is equal to the Doppler shift, Δf.

(e) The filter

This removes any frequencies other than Δf.

(f) The amplifier

The signal is further amplified.

(g) The recorder—analyser

This may be an electrical system for the accurate measurement of Δf (and hence v) or may simply be the ear with the aid of headphones or a loudspeaker. Certain Doppler shifts have a characteristic sound (e.g. the foetal heart movement produces a range of Δf's resembling the sound of galloping horses; the placenta a 'sound' similar to that of wind rushing through trees) and the experienced ear can recognise any abnormalities.

A Doppler system operating at frequencies around 2–3 MHz may be used to investigate the timing and velocity of heart valves. Since the velocity of the valves is considerably greater than that of the surrounding structures, the larger Doppler shifts are easily identified with valve movements. The system is more accurate than those using pulse–echo techniques (page 125).

Blood flow measurement

The Doppler shift technique may be used to estimate the velocity of blood flow in veins and arteries. Since these are generally superficial, attenuation is consequently less important than with deeper structures, and higher ultrasound frequencies in the range 5–10 MHz, are used, giving better resolution.

The ultrasound probe is generally positioned on the body surface so that the ultrasound beam strikes the blood vessel at an angle θ (see Fig. 10.11). The beam is reflected by particles of blood, which have a velocity v relative to the vessel, but which have a component $v \cos \theta$ in the direction of the beam. Thus, the situation is analogous to that of a moving reflector of velocity $v \cos \theta$, and the resulting

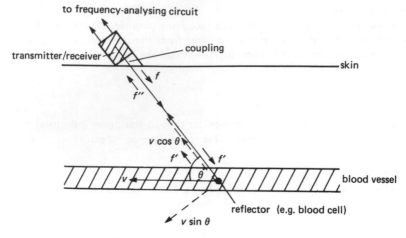

Figure 10.11   Blood flow measurement

Doppler shift is given, using equation [10.3], by:

$$\Delta f_3 = \frac{2fv \cos \theta}{c}$$

If Δf_3 is measured, using the same system as displayed in Fig. 10.10, then, knowing f, c and $\cos \theta$, v may be evaluated.

In practice, since the blood is not all flowing at the same velocity, this being higher in the centre than near the walls of the vessel, various Doppler shifts are detected. The analysis and measurement of these shifts is a complex and difficult procedure.

The main advantage of the system is that surgical penetration of the vessel is unnecessary. Continous recordings can be made with no risk or discomfort to the patient and can lead for example to flow volume estimates if the vessel diameter is known. Thrombosis is easily diagnosed, since any vessel constrictions are readily identified by the change they produce in the normal signal.

One disadvantage of the simple Doppler flowmeter is that it does not differentiate between movement towards and away from the transducer. This is not the case, however, with recently developed (and more expensive!) Doppler systems, which incorporate colour into their displays. The ultrasound "image" of the scanned body section is shown on a CRO, and a range of colours is used to represent the different blood flow velocities. Such colour Doppler imaging of the heart is proving particularly successful, since any abnormal holes, blockages or valve defects are readily identified from the blood flow pattern.

Biological effects of ultrasound

The interaction of ultrasound with biological tissue is either thermal or non-thermal. The thermal mechanism involves the conversion of sonic energy into heat as the ultrasound beam is absorbed in the medium. With the exception of some Doppler systems, the average intensities of diagnostic instruments are small enough to prevent temperature rises in excess of 1 K. However, with the higher intensities ($\approx 10^4 \, \text{W m}^{-2}$) employed in ultrasound therapy, significant heating of deep tissue occurs often with beneficial effects. Bone, having a high absorption coefficient (see Table 10.1) is at risk in ultrasound therapy and careful control of the beam is essential.

Non-thermal effects are not fully understood at present, but are thought to be due mainly to *cavitation*. If a sufficiently intense ultrasound beam encounters gas bubbles in its path, the resulting increase in energetic activity in the vicinity of the bubbles is capable of disrupting cells and tissues. Such effects, though unlikely at diagnostic intensities, occur in ultrasound therapy. Whilst presenting a potential hazard, they are also thought to be responsible for the acceleration of wound healing and bone repair, thus assisting in the treatment of conditions such as arthritis and skin ulcers.

The advantages of ultrasound

The major advantages offered by ultrasound, particularly in preference to radiological techniques, are:

(a) soft tissue structures are easily examined; hence the widespread use of ultrasound in obstetrics and gynaecology;
(b) there are no harmful side effects during diagnostic investigations; low-power ultrasound is very safe and non-accumulative;
(c) The equipment is relatively simple, can be portable and can be operated using an ordinary wall socket.

Exercise 10

1 Explain the following terms:
(a) ultrasound,
(b) acoustic impedance,
(c) intensity reflection coefficient.

Why is a coupling medium necessary between a source of ultrasound and the body during ultrasonic investigation of the body? The velocities of sound in air, oil and tissue (average) are 0.33, 1.50 and 1.53 km s^{-1} respectively. The densities of the three media respectively are 1.3, 950 and 1065 kg m^{-3}. Estimate the percentage of the sound intensity reflected at:
(d) an air–tissue interface:
(e) an oil–tissue interface;
(f) an air–oil interface.
Comment on the relevance of your results.

2 Explain what is meant by piezoelectricity and describe how the principle is applied in an ultrasound generator.

A quartz crystal is used in an ultrasound transducer placed on the body surface. Using the data below, estimate whether transmission of ultrasound into the body would be improved by using an intermediate perspex window. (You may neglect the additional attenuation.)

Acoustic impedance of quartz $= 1.5 \times 10^7 \, \mathrm{kg \, m^{-2} \, s^{-1}}$

Acoustic impedance of perspex $= 3.2 \times 10^6 \, \mathrm{kg \, m^{-2} \, s^{-1}}$

Acoustic impedance of tissue (average) $= 1.63 \times 10^6 \, \mathrm{kg \, m^{-2} \, s^{-1}}$

3 (a) What are the essential features of a system which uses ultrasound to construct images of structures inside the body? Explain how the sound is introduced into the body and which particular physical properties of the tissues govern the quality of the image produced. Suggest a modification of the method which might allow one to observe the speed of movement of a pulsating organ such as the heart.

(b) Mention two advantages that ultrasonic methods have over techniques employing X-rays for the production of images of internal organs. [JMB]

4 Using a simple block diagram, describe the essential components of an ultrasonic A-scan system and briefly indicate their functions. Give two examples of the use of the A-scope in medical diagnosis.

Given that the acoustic impedances of quartz, water, oil and tissue are respectively 1.5×10^7, 1.5×10^6, 1.42×10^6, and $1.63 \times 10^6 \, \mathrm{kg \, m^{-2} \, s^{-1}}$, and neglecting attenuation, calculate the percentage transmission of an ultrasound beam travelling from quartz into tissue via an intermediate medium of:

(a) water,

(b) oil.

Why is oil used in preference to water as a coupling medium in most diagnostic studies?

5 Explain carefully what information is being plotted on the cathode ray tube screen during:

(a) an A-scan,

(b) a B-scan,

(c) a time–position scan.

A beam of ultrasound of frequency 1 MHz passes through 6 mm of fat before striking normally a fat–muscle interface. Using the data from Table 10.1 calculate:

(d) the percentage of the incident intensity arriving at the interface;

(e) the percentage of the original incident intensity transmitted into the muscle.

6 In the following investigations, state which type of scan(s) would be used, and describe clearly how the results would be obtained:

(a) measurement of foetal head size;

(b) diagnosis of twins;

(c) diagnosis of mitral valve stenosis (constriction of one of the heart valves);

(d) measurement of lens-to-retina distance;

(e) localisation of the placenta.

Why is it necessary to 'rock' the ultrasonic probe, as it is moved over the patient during a B-scan?

7 Derive, from first principles, an expression for the Doppler shift in frequency occurring when ultrasound waves of frequency f are transmitted into a medium

in which their velocity is c, and reflected normally from a plane interface moving with a velocity v, (a) directly towards the waves, (b) directly away from the waves.

What implication does this have for a Doppler system which registers only the magnitude of the Doppler shift?

A 2.5 MHz ultrasound beam travels at 1.5 km s^{-1} through soft tissue before being reflected normally from a moving interface. If a Doppler shift of 500 Hz is detected in the reflected beam, estimate the velocity of the moving surface.

8 In a determination of blood flow, the transducer is set at an angle θ to the flow. Show that the recorded shift in frequency is given by:

$$\Delta f = \frac{2fv \cos \theta}{c}$$

where f is the frequency of the transmitted waves, v is the blood velocity, and c is the average velocity of the ultrasound.

Indicate how Δf can be measured in a simple Doppler flowmeter, and mention one advantage and one disadvantage of such a system.

During such a blood flow determination, ultrasound of frequency 3 MHz is transmitted at an angle of 30° to the blood vessel. If the velocity of ultrasound is taken to be 1.55 km s^{-1} and the diameter of the vessel to be 1.2 mm, estimate the blood volume flow rate if a Doppler shift in frequency of 1500 Hz is recorded.

9 (a) With the help of a sketch of a typical transducer describe the generation of ultrasonic waves for medical diagnostic purposes.
   Explain:
   (i) why it is essential to use short pulses of ultrasound;
   (ii) how it is ensured that the sound energy enters the body;
   (iii) the means employed for detection of the received signals.

(b) For pulses of a given power emitted by the transducer mention **two** factors which affect the intensity of the received ultrasound signals.

(c) If the transducer were to be part of a B-scanner, what other fact would you need to know about the received ultrasonic signals? Apart from such data about the received signals, state the additional information needed by the B-scanner if an image is to be produced. [JMB]

PART III
Ionising radiation

11 X-rays

The origin of X-rays

Emission of X-rays

X-radiation is electromagnetic radiation of high quantum energy (greater than about 1 keV). It may be produced by bombarding a metal target with high-speed electrons emitted from a heated cathode by thermionic emission and accelerated towards the positive anode and target (Fig. 11.1).

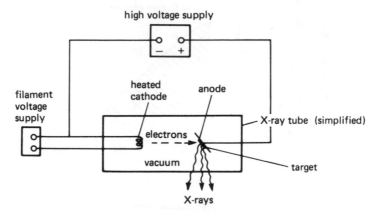

Figure 11.1   Production of X-rays

On striking the target, the electrons lose most (about 99 per cent) of their energy in low-energy collisions with target atoms resulting in a substantial heating of the target. The rest (usually less than 1 per cent) of the electron energy reappears as X-radiation, which generally has the wavelength and energy distributions displayed in Fig. 11.2.

X-ray spectra

The continuous spectrum of Fig. 11.2 is the result of some electrons decelerating rapidly in the target by transferring much of their energy to single photons. This 'braking radiation' or 'bremsstrahlung' has a maximum energy, E_{max}, corresponding to the maximum energy of the bombarding electrons:

$$E_{max} = eV_p$$

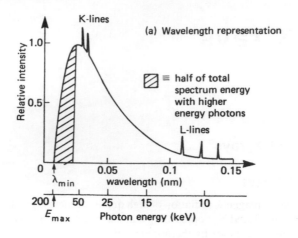

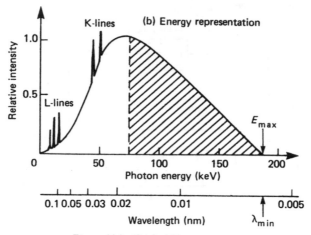

Figure 11.2   Typical X-ray spectra

where V_p is the peak (maximum) voltage across the X-ray tube. The equivalent minimum wavelength λ_{min} of the X-ray emission is thus:

$$\lambda_{min} = \frac{c}{\text{max frequency}} = \frac{ch}{E_{max}} = \frac{ch}{eV_p}$$

Superimposed on the continuous spectrum are a number of sharp intensity peaks constituting a line spectrum. The lines occur in groups, the shortest wavelength (highest energy) group being called the K-lines, the next the L-lines, and so on. These lines are a result of bombarding electrons penetrating deep into target atoms and ejecting orbital electrons from the innermost shells (the K, L, shells, and so on) near the nuclei. Electrons from outer orbits subsequently make transitions to fill the gaps in the inner shells, thereby emitting photons whose energies are characteristic of the target atom. Transitions terminating in the K-shell give rise to the K-lines, those terminating in the L-shell produce the L-lines, and so on. As long as the target has a high enough atomic number, Z, the resulting photon energies will be in the X-ray range. For example, tungsten ($Z = 74$) has K-line wavelengths around 0.02 nm, corresponding to photon energies of about 70 keV. Tube voltages of at least 70 kV are hence required to generate tungsten's characteristic K-lines.

Factors affecting the X-ray beam

Tube voltage

The following effects are noted when tube voltage is increased (Fig. 11.3(a)):

(a) E_{max} ($\propto V_p$) increases; λ_{min} ($\propto 1/V_p$) decreases;

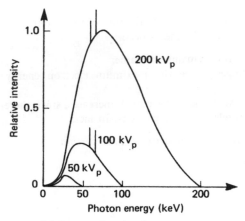

(a) Change of peak voltage

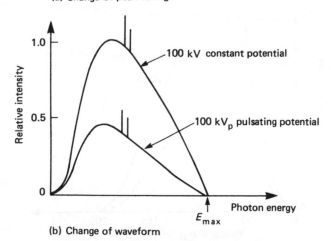

(b) Change of waveform

Figure 11.3 Effect of tube voltage on spectrum

(b) the peak of the continuous spectrum moves towards higher energies;
(c) the total intensity (area under the curve) increases rapidly, since more of the bombarding electrons now have energies sufficient to produce X-rays; in fact:

$$\text{total output intensity} \propto V_p^2$$

(d) more characteristic lines may appear in the spectrum as the minimum wavelength limit decreases.

The tube waveform also determines the shape of the X-ray spectrum, (Fig. 11.3(b)). In practice, the spectrum often lies between the two extremes shown and depends largely on the rectification employed in the high-tension circuit, (see page 153).

Tube current

If the tube current is increased by increasing the rate of thermionic emission from the cathode, (Fig. 11.4) the following effects occur:

(a) the shape of the spectrum remains the same;
(b) E_{max} remains unchanged, since the maximum electron energy available remains unchanged;
(c) the overall intensity (area under the curve), increases, since there are more electrons available to release X-ray photons. In fact:

$$\text{total output intensity} \propto \text{tube current}$$

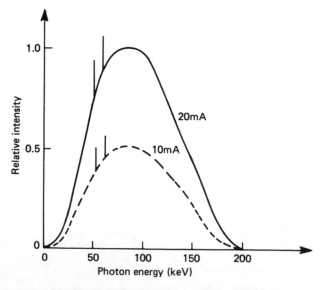

Figure 11.4   Effect of tube current on spectrum

Target material

An increase in target atomic number Z, (Fig. 11.5) results in:

(a) an overall increase in X-ray intensity since the greater mass, size and positive charge of the target nuclei lead to a greater probability that bombarding electrons make collisions resulting in X-ray emission. In fact:

$$\text{total output intensity} \propto Z$$

(b) a shift of the characteristic line spectrum towards higher photon energies since more energy is needed to expel K, L, electrons etc. from higher Z atoms;

(c) no change in E_{max} since the maximum electron energy available remains unchanged.

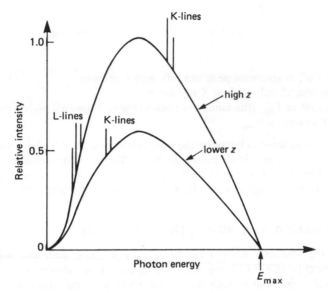

Figure 11.5   Effect of target material on spectrum

A target material should not only have a high melting point but also a high Z, to yield high X-ray outputs. Tungsten, with a melting point of 3650 K and a Z of 74, is almost universally used as a target material.

Filtration

If a sheet of metal or other material is placed in the path of the X-ray beam it acts as a filter, selectively absorbing more lower-energy photons than high-energy photons (see Fig. 11.6). Such filtration thus produces:

(a) a change in X-ray spectrum shape with the preferential removal of lower energies;

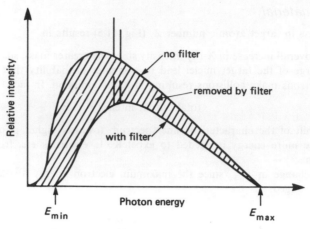

Figure 11.6   Effect of filtration on spectrum

(b) a shift in spectrum peak towards higher energies;
(c) an overall reduction in X-ray output;
(d) a shift in E_{min} (the minimum photon energy) towards higher energies;
(e) no change in E_{max}.

Although reduced in intensity, the transmitted beam is relatively more penetrating than before and is described as being 'harder'. The amount of radiation absorbed by the filter depends on photon energy, filter thickness and filter material (high-density and high-Z materials absorbing the most radiation).

Attenuation and absorption of X-rays

Firstly, the intensity of an X-ray beam may decrease with distance from the source simply due to divergence. For spherical wavefronts emitted from a theoretical point source, such divergence leads to a reduction in intensity described by the inverse square law, ($I \propto 1/r^2$).

Secondly, when the X-ray beam traverses a medium, the photons may be scattered or absorbed in the medium, leading to an overall *attenuation* (reduction in intensity) of the beam.

Attenuation of a parallel homogeneous X-ray beam

A homogeneous, or monoenergetic, X-ray beam contains photons of only one energy. When such a beam passes through a medium, the fractional reduction in intensity $(-dI)/I$ in passing through a small distance dx is proportional to dx.

$$\therefore \qquad \frac{-dI}{I} = \mu dx$$

where μ is a constant known as the total linear attenuation coefficient. Integrating gives:

$$[\log_e I]^I_{I_0} = -[\mu x]^x_0$$

where I_0 is the incident intensity and I is the transmitted intensity after traversing a thickness x of the medium.

$$\therefore \qquad I = I_0 e^{-\mu x} \qquad [11.1]$$

and the transmission falls off exponentially (Fig. 11.7). μ depends on both the medium (being greatest for high Z and high-density materials) and the X-ray beam (being greatest for low photon energies).

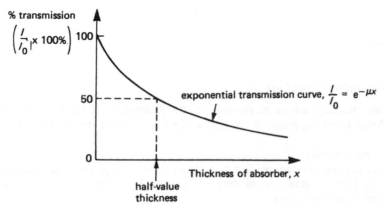

Figure 11.7 Transmission curve for homogeneous radiation

It is often convenient to use a mass attenuation coefficient, μ_m, which describes the attenuation per unit mass of material traversed.

$$\mu_m = \frac{\mu}{\rho}$$

where ρ is the density. For a given Z and photon energy E:

$$\mu \propto \rho \text{ and } \frac{\mu}{\rho} \text{ is constant}$$

thus giving μ_m a particular value which is independent of density. Therefore, μ_m depends only on Z and E (see Table 11.1, on page 146).

Half-value thickness

The penetrating power, or quality, of a homogeneous beam of X-rays, (or γ-rays), can be described in terms of the half-value thickness, HVT, (or half-value layer, HVL) in a given material. This is that thickness of the material which reduces the intensity of a narrow homogeneous beam to half its original value. Thus, putting

$I = I_0/2$ and $x = x_{1/2}$ (HVT) in equation [11.1] gives:

$$\log_e \tfrac{1}{2} = -\mu x_{1/2}$$

∴
$$x_{1/2} = \frac{\log_e 2}{\mu} = \frac{0.693}{\mu}.$$

For example, a beam of 80 keV X-ray photons has a HVT of 1 mm in copper, whilst a beam of 1 MeV γ-ray photons has a HVT of 10 mm in lead.

The HVT at a given photon energy in a particular material may be determined experimentally by placing various thicknesses of the material in the path of a narrow beam of the appropriate energy and measuring the transmitted intensity, for example, using a thimble ionisation chamber (see page 184). Since

$$\log_e I - \log_e I_0 = -\mu x$$

a plot of $\log_e I$ against x yields a straight line of slope $-\mu$. From this $x_{1/2}$ can be calculated.

Attenuation mechanisms

Various mechanisms are responsible for the attenuation of an X-ray beam in a medium. Those important in soft tissue (see Fig. 11.8) are described below.

(a) Simple scatter
The incident photon energy (E) is much less than the energy required to remove an electron from its atom (the binding energy, E_b). The photon is simply deflected without change of energy.

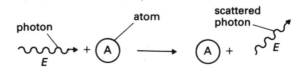

photon

atom

scattered photon

E

E

Figure 11.8 (a) Simple scatter

(b) Photoelectric effect
The photon, (E slightly greater than E_b), gives up all of its energy to an inner orbital electron, thereby ejecting it from its atom. The excited atom subsequently returns to its ground state with the emission of characteristic photons, most of which are

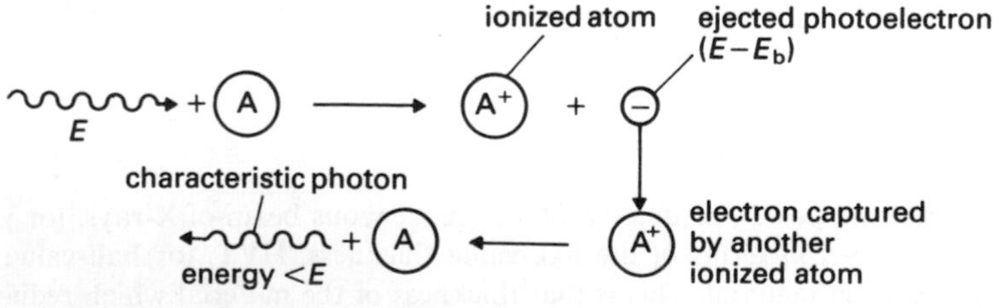

ionized atom   ejected photoelectron ($E-E_b$)

E

characteristic photon

energy $<E$

electron captured by another ionized atom

Figure 11.8 (b) Photoelectric effect

of relatively low energy and are immediately absorbed in the material itself. The ejected photoelectron ionises further atoms along its path until its (often considerable) kinetic energy has been dissipated.

(c) Compton scatter

E is much greater than E_b, and only part of the photon energy (E') is given up during the interaction with an outer (valence) electron which is effectively 'free'. The photon continues in a different direction with diminished energy, $(E - E')$, and the electron (known as a recoil electron) dissipates its energy through ionisation.

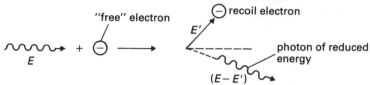

Figure 11.8 (c) Compton scatter

(d) Pair production

At very high photon energies, the incident photon approaches and interacts with the nucleus of an absorber atom, with the result that the photon disappears and two particles (an electron and a similar but positively-charged particle, the positron) emerge. Since the mass of each particle is 9.11×10^{-31} kg, the minimum photon energy necessary for such pair production may be estimated using Einstein's relation:

$$E = mc^2$$
$$E_{min} = 2 \times 9.11 \times 10^{-31} \times (3 \times 10^8)^2 \text{ J}$$
$$= 1.64 \times 10^{-13} \text{ J}$$
$$= 1.02 \text{ MeV}$$

The two particles continue through the absorber, losing energy by ionisation until electron–positron annihilation occurs, yielding two photons, each of energy 0.51 MeV, moving in opposite directions to conserve momentum.

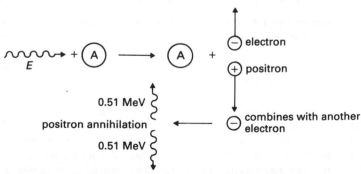

Figure 11.8 (d) Pair production and positron annihilation

Table 11.1 Attenuation mechanisms

Mechanism	Variation of μ_m with E (photon energy)	Variation of μ_m with Z (atomic number)	Energy range in which dominant in soft tissue
Simple scatter	$\propto 1/E$	$\propto Z^2$	1–20 keV
Photoelectric effect	$\propto 1/E^3$	$\propto Z^3$	1–30 keV
Compton scatter	falls very gradually with E	independent	30 keV–20 MeV
Pair production	rises slowly with E	$\propto Z^2$	Above 20 MeV

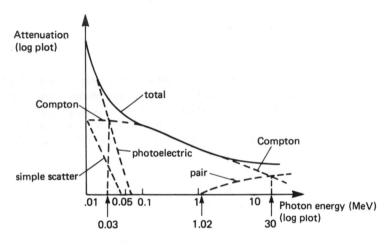

Figure 11.9   Attenuation mechanisms in water

Relative importance of the attenuation mechanisms

Table 11.1 summarises the important features of the attenuation mechanisms including the variations of their mass attenuation coefficients, μ_m. Figure 11.9 illustrates their relative contribution to total attenuation in water which is similar to soft tissue. The general trend of the graph shows that as photon energy is increased (higher X-ray tube voltage) total attenuation diminishes, thus the radiation is relatively more penetrating.

The optimum photon energy for radiography is around 30 keV, (tube voltage ≈ 80–100 kV$_p$). At this energy, the photoelectric effect dominates the attenuation in bone. The variation of this attenuation with Z^3 means that different materials are easily distinguished. For example, the average values of Z for fat, muscle and bone are 5.9, 7.4 and 13.9 respectively, leading to a photoelectric attenuation in bone about eleven times greater than that in surrounding tissue. On the other hand,

Compton scatter is important in soft tissue at this energy, whilst the contribution from simple scatter, which has an adverse effect on film quality, is small.

In contrast, deep therapy should be avoided in this energy range, because of the potential danger to bone. Energies above 30 keV (and more commonly in the range 0.5–5 MeV) are hence selected for radiotherapy, so that Compton scatter (with its independence of Z) is the dominant attenuation mechanism.

X-ray filters

The primary beam of X-rays emitted from an X-ray tube passes through the glass wall of the tube, a layer of oil and a shield aperture before emerging, so that some absorption of the lower-energy photons always takes place. For example, this inherent filtration of a diagnostic tube may be equivalent to that produced by about 0.5–1.0 mm of aluminium.

Despite inherent filtration, the X-ray beam still contains a high proportion of low-energy radiation, which simply becomes absorbed in the patient's skin. Such radiation (except in the radiotherapy treatment of superficial areas) only adds unnecessarily to the patient's absorbed dose, and is usually removed using additional filtration, thereby hardening the beam.

Suitable filters consist of high-Z materials, in which photoelectric attenuation dominates in the given energy range, since this then results in a selective absorption of low-energy photons. In diagnostic radiography and superficial radiotherapy (tube voltage, $V_p \approx 120$ kV) filters up to a few millimetres thick of aluminium are commonly used. For the higher energies employed in deep therapy aluminium absorbs too much by Compton scatter and higher-Z filters are used of copper ($V_p \approx 120$–200 kV), tin ($V_p \approx 200$–400 kV), lead ($V_p \approx 800$–2000 kV), or gold (V_p above ≈ 2000 kV). Often, composite filters are employed so that the required attenuation is achieved and any characteristic radiation emitted in one material is absorbed in the other.

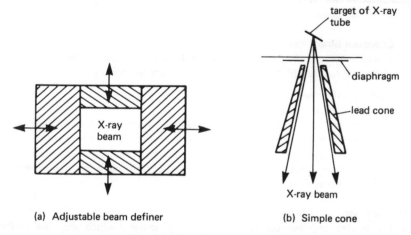

(a) Adjustable beam definer   (b) Simple cone

Figure 11.10   Simple beam definers

Beam size

The cross-section of an X-ray beam may be determined by a beam definer or diaphragm. Figure 11.10(a) illustrates an adjustable diaphragm, consisting of two pairs of metal sheets, which can move at right angles to each other, so that a rectangular cross-section can be defined. Another simple beam definer is a section of a cone as shown in Fig. 11.10(b).

Such designs often incorporate a light box to assist with alignment of the beam. A lamp and mirror are positioned so that a light beam of exactly the same size, shape and direction as the X-ray beam illuminates the patient.

Narrow beams are preferred in radiography, since the random scatter which inevitably increases with wider beams blurs the radiographic image. On the other hand, wide beams can be advantageous in radiotherapy, particularly for the treatment of deep sites where the extra scatter is useful. Therapy demands a sharp edge of the beam at the patient and extra collimators are then used.

The radiographic image

Factors affecting image quality

On a clear radiograph, regions of high attenuation (bone) appear white, medium attenuation (tissue) appear grey and negligible attenuation (air) appear black. In practice, there are many factors, some conflicting, which influence the quality of the radiographic image formed after an X-ray beam traverses the body.

(a) Tube voltage
This is generally in the range 60–$125\,kV_p$, providing photon energies around $30\,keV$, the optimum value for good contrast. (See page 146).

(b) Tube current
An increase in this results in an overall increase in intensity of both incident and transmitted beams but at the expense of larger patient-absorbed doses and increased heating of the target.

(c) Exposure time
This determines the density or blackening of the film and although longer exposure times improve radiographic contrast, such times are limited by:

  (i) patient-absorbed dose;
 (ii) overheating of the target;
(iii) movement blur or unsharpness, which is produced by movement of the examined structure during exposure. Whereas the limbs, for example, can be immobilised to allow exposure times of a few seconds, organs like the stomach are subject to involuntary movements and permit maximum exposure times of only about 0.5 s.

(d) Beam size
Narrow beams improve contrast by reducing scatter, which may be further eliminated using special grids, (see page 150).

(e) Focal spot size

A small focal spot produces sharper images but a greater concentration of heat in the target (see below).

(f) Filtration

This reduces unwanted low-energy radiation and scatter giving some improvement in contrast.

(g) Artificial contrast media

If the natural contrast offered by the relevant body parts under study is insufficient, artificial contrast agents may be used. For example, the radio-opaque barium sulphate in an aqueous suspension is widely administered for gastrointestinal tract examinations.

(h) Intensifying screens

Their use increases film density (see page 151).

(i) Detectors

All detectors introduce some blur, known as detector blur, around the final image. The amount of detector blur is small for direct exposure films (<0.1 mm), increases with the use of intensifying screens (≈ 0.25–0.5 mm), and is greatest for fluoroscopic screens (≈ 1 mm) for which the resolution is consequently poorest.

Size of focal spot

X-rays originate from a small area of the target known as the focus or focal spot (see Fig. 11.11(a)). When a radio-opaque object is positioned in the resulting beam, its 'shadow' is cast on the detecting photographic film beyond. A theoretical point focus would produce a well-defined region of full shadow or umbra on the film,

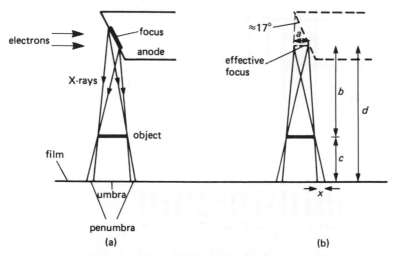

Figure 11.11   Effect of focal-spot size on radiographic image

thus forming a clear image. However, in practice the focus is of a finite size and each point on the focus produces its own slightly different image on the film. The result is a central umbra surrounded by a blurred region of partial shadow or penumbra.

The width x of this penumbra blur (Fig. 11.11(b)) is called the geometric unsharpness of the image and can be estimated by considering an effective square focus of side a. This is the projection of the real focus on to a plane at right angles to the beam direction. Using the notation of the figure, by similar triangles:

$$\frac{x}{a}=\frac{c}{b}=\frac{c}{d-c}$$

$$\therefore \quad x=\frac{ac}{d-c}$$

Hence, to reduce penumbra blur, a and c should be small and d large.

Reduction of scatter

To reduce the blurring of images caused by scattered radiation, a grid may be used directly in front of the detector (Fig. 11.12). The grid consists of a large number of lead strips about 0.05 mm thick and 5 mm long, held 0.5 mm apart by an interspace material transparent to X-radiation. Direct, or primary radiation reaches the film through the gaps whereas scattered radiation is intercepted by the grid. If two 'crossed grids' are used at right angles to each other almost complete elimination of unwanted scatter occurs.

The formation of an image of the lead strips on the film, sometimes referred to as grid lines, may be prevented by moving the grid slowly across the film during exposure.

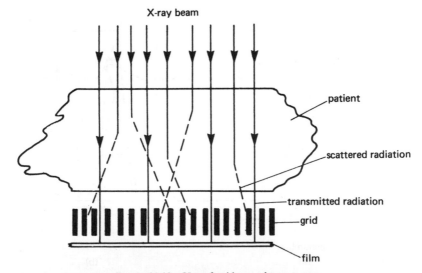

Figure 11.12   Use of grid to reduce scatter

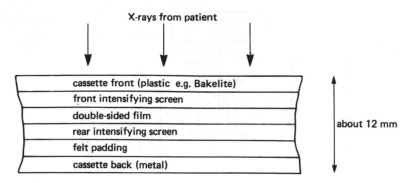

Figure 11.13 An intensifying screen cassette

The intensifying screen

Intensifying screens are used to increase the density on the photographic film thus permitting shorter exposure times. They employ fluorescent screens which absorb the X-radiation and re-emit visible radiation in a pattern duplicating that of the original X-ray beam.

In the simple intensifying screen cassette (Fig. 11.13) two fluorescent screens (commonly zinc sulphide crystals bonded to cardboard) are placed in intimate contact with the two faces of a light-sensitive X-ray film having thin emulsions on both sides. A layer of felt padding helps to ensure good screen–film contact, and the whole is contained in a light-proof plastic envelope called a cassette. The metal backing to the cassette provides some radiation protection for operational personnel and reduces scatter back into the film 'sandwich'.

X-rays from the patient penetrate the film and two fluorescent screens. Although a slight latent image is thus directly produced on the two film emulsions, their major activation results from the light produced in the adjacent fluorescent screens. In practice, about 97 per cent of the final image is due to light from the intensifying screens, and a cassette of this type can have an intensification factor of about 10–40, depending on the type of fluorescent screen used. Although some detail is lost due to the inevitable diffusion of the fluorescent light, exposure times can be reduced by this same factor of about 10–40 which more than compensates for this loss.

The production of X-rays

The X-ray machine (or generator)

The machine (Fig. 11.14) should be adaptable to many different diagnostic and therapeutic applications, and must therefore provide independent control of:

(a) tube current, determined by filament current;
(b) tube voltage; the very wide voltage range required, approximately 0.02–42 MV, is provided by the high-tension circuit;
(c) exposure time.

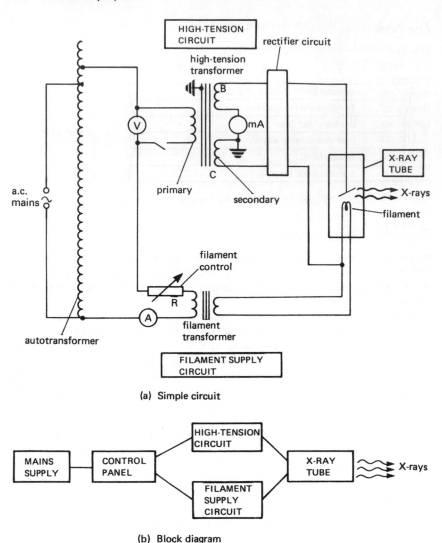

(a) Simple circuit

(b) Block diagram

Fig. 11.14 The X-ray machine (or generator)

Suitable values are set using selector switches on the control panel which also has other meters and safety controls.

The filament supply circuit

This circuit consists essentially of a step-down transformer supplying a relatively low voltage of about 10 V to the tube filament. The filament current ($\approx 4A$) may be adjusted by means of the rheostat R. The ammeter A passes a current related to the filament current and may be calibrated to read filament current directly.

The high-tension circuit

(a) The high-tension transformer

This supplies the high voltage. The primary winding of this step-up transformer consists of a few hundred turns of thick wire carrying a high primary current which may reach about 300 A. The secondary winding of about 100 000 turns of thin, insulated copper wire is wound on an insulating cylinder which is fitted over the primary winding.

The danger of insulation breakdown occurring, due to the high voltages developed across the ends of the secondary, is reduced by earthing the centre of the winding. For example, if $100\,kV_p$ exists across the winding, the ends B and C will only be at a maximum of $\pm 50\,kV_p$ with respect to the earthed transformer core and insulating oil tank in which it is immersed. In addition, the milliammeter mA for monitoring the tube current can be inserted as shown at this earth connection, thereby reducing the electrical hazard of using the meter at a high voltage point elsewhere in the circuit. Using this arrangement, the meter may be safely mounted on the control panel.

Again because of insulation problems, it is difficult to provide a variable high voltage output by varying the turns ratio of the transformer. Instead, the input (primary) voltage is adjusted using an autotransformer, and the X-ray tube voltage, being a multiple of this primary voltage, is monitored using a suitably calibrated voltmeter V in the primary circuit.

(b) The rectifier circuit

This modifies the shape of the voltage waveform, (Fig. 11.15).

   (i) In small portable diagnostic or dental units, where small tube currents and voltages are adequate, the X-ray tube itself acts as a simple rectifier passing current only from filament to anode (Fig. 11.15(a)). However, such self-rectification breaks down at higher tube currents, when increased heating of the target results in thermionic emission here during the reverse voltage stage of each cycle with consequent damage to the tube.

  (ii) A separate rectifier circuit is then commonly employed between the transformer output and the X-ray tube. The half-wave rectified circuit (Fig. 11.15(b)) protects the tube from the reverse voltage but only provides current for half the time, thus limiting X-ray output.

 (iii) The full-wave rectified circuit (Fig. 11.15(c)) provides unidirectional, but not constant, tube current.

 (iv) Further smoothing, using a capacitor (sometimes with an additional smoothing filter), can produce approximately constant currents, (Fig. 11.15(d)).

The X-ray tube

In a typical diagnostic tube (Fig. 11.16) the filament is a flat spiral of heavy tungsten wire, much longer than it is wide. It is heated to incandescence by the variable filament current which controls its temperature, the rate of thermionic emission and hence the tube current. Surrounding the filament is a metal cup, maintained at a high negative potential which serves to converge the emerging electrons towards a focus at the target. Electrostatic focusing of the electron beam is also sometimes employed.

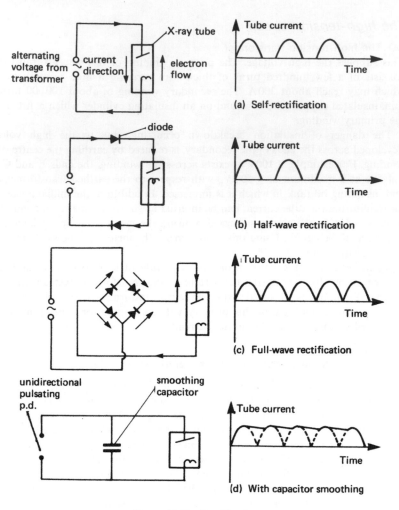

(a) Self-rectification

(b) Half-wave rectification

(c) Full-wave rectification

(d) With capacitor smoothing

Figure 11.15   Rectification

A compound anode is frequently used and consists of a small tungsten target embedded in a large copper anode. Tungsten is a good target material (see page 141) and copper an ideal housing medium with its high thermal conductivity and specific heat capacity. Heat is thus conducted away quickly and excessive anode temperature rises are prevented. In deep therapy X-ray tubes, the copper anode is mounted on a thick but hollow copper stem through which oil is circulated as a coolant. The oil is in turn cooled by water in a heat exchanger.

In most modern diagnostic tubes, a rotating anode (Fig. 11.16) is used. This consists of a solid tungsten disc about 100 mm in diameter with a mass of about 0.5 kg. The periphery is bevelled to an angle of approximately 17° and forms the target surface on to which the electron beam is focused. Part of the anode system is the rotor of an induction motor, by means of which the anode is rotated rapidly

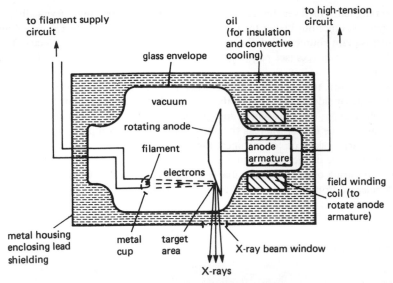

Figure 11.16   A rotating-anode X-ray tube

and smoothly at about 50 revolutions per second on ball-bearings. The target area under electron bombardment is constantly changing and the consequent reduction in local heat concentration permits larger X-ray outputs. Cooling is mainly by heat radiation through the vacuum to the glass wall.

The line focus principle (Fig. 11.17) is applied in all diagnostic and some therapy tubes. A rectangular filament results in a target focal spot, or real focus, which is

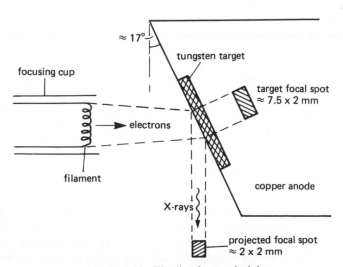

Figure 11.17   The line focus principle

also rectangular. However, by inclining the target at a sharp angle (typically 17° for diagnostic tubes but higher for therapy tubes) the X-rays appear to come from a smaller square source called the projected focal spot, or apparent focus. This provides sharper images (see page 149), without excessive concentration of heating in the target.

Most diagnostic tubes are dual-focus tubes, having two different-sized filaments mounted side by side for selection at the control panel. The large filament, coupled with a short exposure time, suits radiographic examination of regions prone to movement (e.g. the trunk) whereas the small filament with a longer exposure time produces sharp images of, for instance, the limbs.

The high degree of vacuum attainable in modern tubes (10^{-7}–10^{-8} mmHg) may deteriorate during operation, particularly if high temperatures are incurred as gas may be released from the tube components into the 'vacuum'. The tube is then said to be 'gassy' and is rendered useless.

Medical applications

Diagnosis

When investigating structures such as bones the natural 'subject contrast' is sufficient to produce clear radiographs and permit diagnosis of fractures, dislocations, and so on. Other materials including tissues, tumours and blood vessels are not so easily investigated and for these artificial contrast media are often used. Such improved techniques have enabled blood vessel blockages, ruptured spinal discs, tumours and cysts in the spine, and stones in the bile duct to be located, as well as permitting accurate 'visualisation' of the heart.

Moving structures, such as the heart and gastrointestinal tract, are best investigated using fluoroscopy in which the X-ray image is displayed continuously on a fluorescent screen (see page 106). However, the detail and contrast achievable during fluoroscopy are inferior to those obtained in radiography, and the larger radiation doses delivered to the patient during fluoroscopy further restricts its routine use.

Therapy

Radiation therapy generally aims to destroy malignant (cancer) cells in the body without harming healthy tissue or giving discomfort to the patient. Treatment may be in a single dose or, more commonly, in a course spanning several days or weeks, when there is then less sacrifice of healthy tissue.

In order to minimise the damage to intervening healthy tissue (which is fortunately more radiation-resistant than cancer cells) multiple-beam therapy or rotational therapy are usually employed when treating deep tumours. Multiple beams are aimed at the tumour from several directions, making the cumulative effect at the tumour considerable, whilst leaving intervening tissue relatively unharmed. A similar effect is achieved by rotating the patient whilst aiming a single beam at the tumour. Cancers of the pelvis, cervix, larynx and pituitary gland have been successfully treated using these techniques.

Table 11.2 Therapeutic radiations

Treatment	Voltage range	Typical photon energy (MeV)
Superficial therapy	20–150 kV$_p$	0.03
Orthovoltage therapy	140–300 kV$_p$	0.1
Teletherapy:		
^{137}Cs		0.66 (γ)
^{60}Co		1.17, 1.33 (γ)
Supervoltage therapy:		
Linear accelerator	4–8 MV	2
Betatron	9–42 MV	8

Radiations used in therapy

Table 11.2 lists the radiations commonly used in therapy. Teletherapy utilises high-energy γ-radiation, which only differs from equal energy X-radiation in its method of production.

A radioactive source suitable for use in teletherapy should have:

(a) a long half-life to avoid frequent renewal;
(b) a high specific activity (see page 163), so that it may be small;
(c) characteristic γ-radiation of a suitably high energy, around 1 MeV.

^{60}Co has a half-life of 5.3 years and emits γ-rays of the required energy, so is commonly used, (the 'cobalt bomb').

In a ^{60}Co unit, the ^{60}Co source is located near the centre of a lead-filled steel container known as the head. During therapy, an electric motor moves the source so that it comes opposite a shuttered opening in the head, and the emerging γ-rays are collimated before striking the patient. After treatment, the motor once again returns the source to its original location.

Advantages of teletherapy include a constant output (except for the predictable reduction due to radioactive decay), relative simplicity of equipment and lack of problems associated with very high voltages. Set against these is the fact that the source can never be 'turned off' and hence presents a permanent potential radiation hazard. Adequate protection for personnel working in the treatment room must be provided.

Teletherapy and supervoltage both offer several advantages over orthovoltage techniques, one of the most important being the reduction in skin damage. However, supervoltage X-rays are very penetrating and difficult to contain. Great thicknesses (≈ 0.15 m lead) are needed in collimators and other equipment, which presents manipulation problems.

Exercise 11

1 An X-ray tube with a tungsten target is operated at a peak voltage of 80 kV$_p$, without additional filtration. Show graphically how the intensity of the resulting

X-radiation varies with photon energy, and explain the main features of the graph.

How does the graph change if:
(a) the tube voltage is increased to $150\,kV_p$;
(b) an aluminium filter is inserted in the beam?
What is meant by the 'hardening' of an X-ray beam?

2 Explain carefully the effects of:
(a) tube voltage,
(b) tube current,
(c) target material,
on the X-ray spectrum derived from an X-ray tube.

Calculate the maximum photon energy and minimum wavelength of the radiation obtained from an X-ray tube operating at a peak voltage of $70\,kV_p$. ($e = 1.6 \times 10^{-19}\,C$; $c = 3 \times 10^8\,m\,s^{-1}$; $h = 6.6 \times 10^{-34}\,J\,s$).

3 Explain the various mechanisms responsible for the attenuation of an X-ray beam in the body. Discuss the relative importance of these mechanisms:
(a) for beams of different energies;
(b) in different structures of the body.
   What implications are there concerning the most suitable photon energies for use in:
   (i) radiography;
   (ii) radiotherapy?

4 Describe the two main processes by which a beam of X-rays from a tube operated between $30\,kV_p$ and $250\,kV_p$ is attenuated by a thin slab of material. Discuss how the magnitude of each depends on:
(a) the nature of the material;
(b) the quality of the radiation.
   How do such considerations affect the choice of filter material employed in radiology?

5 Describe how the quality of a radiographic image depends on:
(a) X-ray tube current;
(b) exposure time;
(c) size of the focal spot.
   When attempting to improve image quality, are any of these factors conflicting?

   It is required to radiograph a certain structure in the body. The subject is positioned so that the distance from the structure of interest to the film is 0.15 m. If the distance from the film to the target of the X-ray tube is 1 m, and if an effective focal spot 2 mm square is used, estimate a value for the penumbra blur appearing on the radiograph.

6 Why is it important to have independent control of tube current, tube voltage and exposure time in radiology? How is this achieved in a modern X-ray machine?

An X-ray tube is operated with a steady tube voltage of 100 kV and a tube current of 60 mA. Estimate:
(a) the number of electrons hitting the anode per second;
(b) the rate of production of heat at the anode, assuming an efficiency of X-ray production of 1 per cent.
   $(e = 1.6 \times 10^{-19} \text{ C})$.

7 Describe a modern form of X-ray tube and explain its action. What particular features help to prevent overheating of the target?
   An X-ray tube operated at a steady tube voltage of 80 kV and tube current of 150 mA has an efficiency of X-ray production of 1 per cent. Find the intensity of its X-ray beam of cross-sectional area 4 mm²:
(a) emerging from the tube, (assuming no inherent filtration);
(b) after traversing 3 mm of aluminium if the half-value thickness for the beam in aluminium is 1.5 mm.

8 (a) Discuss the use of X- and γ-rays in therapy, explaining carefully why different radiations are chosen for different treatments.
  (b) Describe the essential differences between X-ray units designed for diagnosis and those designed for therapy.

9 (a) (i) Sketch a diagram of a typical diagnostic X-ray tube, identifying appropriate parts of the sketch.
      (ii) Draw a typical X-ray spectrum generated by such a tube and explain how the intensity of the X-rays may be varied.
  (b) (i) Define *half-value thickness* of a material used as an X-ray absorber.
      (ii) Explain why, in producing a radiograph, it is usual to filter the X-ray beam, and suggest a suitable filter material.
      (iii) Identify the process chiefly responsible for the absorption of X-rays by the filter material used in radiography.   [JMB]

10 Explain the following.
  (a) When lifting objects from the ground, it is preferable to bend the knees and keep the back as vertical as possible rather than keep the legs straight and bend forward from the hips.
      Provide a diagram to support your explanation.
  (b) Ultrasound gives images of interfaces between various types of soft tissue even if their densities are very similar.
  (c) Improved contrast in diagnostic X-ray pictures may be obtained by use of a grid.   [JMB]

11 (a) The diagram (Fig. 11.18) shows the relative intensity, I, of an X-ray beam as a function of X-ray photon energy, E, after the beam has passed through an aluminium filter. For one curve the filter thickness was 3.0 mm, while for the other curve the thickness was 1.0 mm.
      (i) Does curve A or curve B correspond to that for the 3.0 mm aluminium filter? Explain your answer by describing the essential differences between curve A and curve B.
      (ii) Why do both curves have the same maximum value of E?
      (iii) Why is the correct filtration of an X-ray beam of importance in diagnostic radiography?

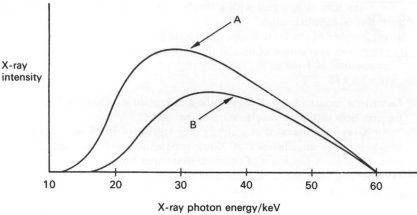

X-ray
intensity

10   20   30   40   50   60

X-ray photon energy/keV

Figure 11.18

(iv) The intensity of a monochromatic beam of X-rays is reduced to $\frac{1}{8}$ of its initial incident value after passing through 3.0 mm of copper. What is the half-value thickness of copper for this radiation?

(b) When sound is transmitted from one medium (medium 1) to another (medium 2), the ratio, R, of the reflected intensity to the incident intensity is given by

$$R = \left(\frac{Z_2 - Z_1}{Z_2 + Z_1}\right)^2$$

where Z_2 is the acoustic impedance of medium 2 and Z_1 is that of medium 1.

In the use of ultrasound for medical diagnosis, a coupling medium such as a water-based cellulose jelly is used between the ultrasonic transducer and the patient's skin. Explain why this is so.

(Acoustic impedance of air $= 0.430 \times 10^3 \, kg \, m^{-2} \, s^{-1}$,
acoustic impedance of water-based jelly $= 1500 \times 10^3 \, kg \, m^{-2} \, s^{-1}$,
acoustic impedance of tissue $= 1630 \times 10^3 \, kg \, m^{-2} \, s^{-1}$.)

Explain the basic principles behind the ultrasound method of obtaining diagnostic information about the depths of structures within a patient's body. Illustrate your answer by reference to a block diagram of a simple (A-scan) scanning system.

What factors limit the quality of the information obtained?   [L]

12 | Production and properties of radioactivity

Notation

The conventional way to represent an atom of element X is $_Z^A$X, where:

$A =$ mass number = number of nucleons (neutrons and protons) in the nucleus

$Z =$ atomic number = number of protons in the nucleus.

A particular variety of nucleus characterised by a specific A and Z, is known as a nuclide, and a series of nuclides having the same Z but varying values of A are known as isotopes of an element.

Radioactive decay

Many elements have several naturally-occurring isotopes but only a limited number are stable. The rest, because their neutron–proton ratio is beyond certain limits, are unstable and decay to a stable form by emission of particles and photons from the nucleus. Such radioactive decay, emission or disintegration, can result in a change in A and Z of the nuclide, which is known as a radionuclide or radioisotope. In addition to the naturally-occurring radioisotopes, there are also many artificially-produced ones.

All radioactive decays follow the same statistical law of radioactivity, namely that in a sample of particular radioisotope the number of atoms which disintegrate in unit time is directly proportional to the number of radioactive atoms present in the sample at that time, that is:

$$-\frac{\mathrm{d}N}{\mathrm{d}t} \propto N$$

or
$$\frac{\mathrm{d}N}{\mathrm{d}t} = -\lambda N$$

where N is the number of atoms remaining and λ is a constant, known as the decay constant, characteristic of the particular radioisotope concerned. If N_0 is the number of radioactive atoms present at time $t = 0$ and N_t is the number remaining after a time t, integrating gives:

$$\int_{N_0}^{N_t} \frac{\mathrm{d}N}{N} = -\lambda \int_0^t \mathrm{d}t$$

$$\therefore \log_e\left(\frac{N_t}{N_0}\right) = -\lambda t$$

$$\therefore N_t = N_0 e^{-\lambda t} \qquad [12.1]$$

Thus, a radioisotope decays exponentially as shown in Fig. 12.1.

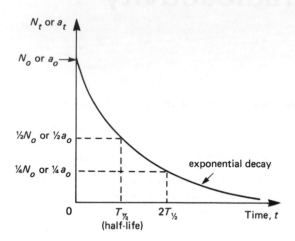

Figure 12.1   Radioactive decay

Activity and half-life

Activity

The activity of a quantity of a radioactive nuclide describes the rate at which disintegrations are occurring. If dN is the number of spontaneous nuclear transformations occurring in the time interval dt, then the activity a is defined by

$$a = \frac{dN}{dt}$$

The unit of activity is hence second^{-1}, otherwise known as the becquerel (Bq); thus:

$$1s^{-1} = 1\,Bq$$

If a_0 is the activity at time $t=0$ and a_t is the activity after a time t:

$$a_0 = -\left(\frac{dN}{dt}\right)_0 = \lambda N_0$$

and

$$a_t = -\left(\frac{dN}{dt}\right)_t = \lambda N_t$$

The minus signs are included to make the activity positive, since dN, representing a decrease in N, is negative.

$$\therefore \frac{a_t}{a_0} = \frac{N_t}{N_0} = e^{-\lambda t}$$

$$\therefore a_t = a_0 e^{-\lambda t} \qquad\qquad [12.2]$$

and a also falls exponentially (Fig. 12.1). The term $e^{-\lambda t}$ is known as the decay factor and is the fraction of the initial activity remaining after a time t.

Specific activity

An important term in the preparation of a sample is its specific activity, this being defined as the activity divided by the total mass of the element present whose radioisotope is involved. Inactive isotopes of the same element behave chemically the same as the radioisotope itself and are referred to as carriers. If no such carriers are present, then the sample is described as being carrier-free.

Half-life

The half-life, $T_{\frac{1}{2}}$, of a radioisotope is defined as the time taken for the activity to fall to half its initial value, or alternatively as the time taken for half the radioactive atoms to disintegrate (see Fig. 12.1).

$$a_t = a_0 e^{-\lambda t} \qquad \text{or} \qquad N_t = N_0 e^{-\lambda t}$$

$$\tfrac{1}{2}a_0 = a_0 e^{-\lambda T_{\frac{1}{2}}} \qquad \text{or} \qquad \tfrac{1}{2}N_0 = N_0 e^{-\lambda T_{\frac{1}{2}}}$$

$$\therefore \lambda T_{\frac{1}{2}} = \log_e 2 = 0.693$$

$$\therefore T_{\frac{1}{2}} = \frac{0.693}{\lambda} \qquad\qquad [12.3]$$

The half-life of a radioisotope is a constant describing how fast the material is disintegrating. It is closely linked with the radioisotope's activity, since:

$$a = -\left(\frac{dN}{dt}\right) = \lambda N$$

$$= N \times \frac{0.693}{T_{\frac{1}{2}}} \qquad\qquad [12.4]$$

The number N of radioactive atoms present per kilogram of the radioisotope, can be evaluated using Avogadro's number:

$$N \text{ per kg} = \frac{6.02 \times 10^{26}}{A}$$

where A is the radioisotope's mass number. Substituting this expression into equation [12.4] gives the activity per kilogram of a carrier-free radioisotope as:

$$a = \frac{6.02 \times 10^{26} \times 0.693}{A T_{\frac{1}{2}}}$$

and the activities of different carrier-free radioisotopes may easily be compared.

For example, if the activity of 1 kg of radium (^{226}Ra) is 3.7×10^{13} Bq and its half-life is 1600 years, then the activity of 1 kg of phosphorus (^{32}P) of half-life 14.3 days is found using:

$$\frac{(a)_\text{P}}{(a)_\text{Ra}} = \frac{(AT_{\frac{1}{2}})_\text{Ra}}{(AT_{\frac{1}{2}})_\text{P}}$$

$$\therefore \frac{(a)_\text{P}}{3.7 \times 10^{13}} = \frac{226 \times 1600 \times 365}{32 \times 14.3}$$

$$\therefore (a)_\text{P} = 1.07 \times 10^{19} \text{ Bq}$$

Table 12.1 lists some common radioisotopes with their half-lives, and illustrates that the naturally-occurring radioisotopes in evidence today have very long half-lives.

Table 12.1 The half-lives of various radioisotopes

Radioisotope	Element	Half-Life
Natural		
^{3}H (Tritium)	Hydrogen	12–26 years
^{14}C	Carbon	5760 years
^{40}K	Potassium	1300 million years
^{226}Ra	Radium	1600 years
^{238}U	Uranium	4500 million years
Artificial		
^{99m}Tc	Technetium	6 hours
^{24}Na	Sodium	15 hours
^{32}P	Phosphorus	14.3 days
^{60}Co	Cobalt	5.3 years
^{125}I	Iodine	60 days
^{131}I	Iodine	8 days
^{137}Cs	Caesium	33 years

Physical, biological and effective half-life

Once a radioisotope has been administered to a patient, it is subject to biological removal from the body by processes such as respiration, urination and defaecation. This means that its effective half-life T_E is less than the physical half-life T_R due to pure radioactive decay. It is possible to define a biological half-life T_B as the time taken for biological processes to remove half the available material, assuming that no new material is arriving.

If λ_B and λ_R are the fractions of the radioisotope removed per second by biological processes and radioactive decay respectively, then the total fraction removed per second is:

$$\lambda_\text{E} = \lambda_\text{B} + \lambda_\text{R} \qquad\qquad [12.5]$$

λ_E is the effective decay constant, and is related to the effective half-life by:

$$\lambda_E = \frac{0.693}{T_E}$$

λ_B and λ_R are similarly related to T_B and T_R. Hence, substitution in equation [12.5] gives:

$$\frac{1}{T_E} = \frac{1}{T_B} + \frac{1}{T_R}$$

For example, the human serum albumin labelled with ^{131}I is often administered. ^{131}I has a physical half-life of 8 days, but is cleared from the body with a half-life of 21 days. The effective half-life is thus calculated using

$$\frac{1}{T_E} = \frac{1}{8} + \frac{1}{21}$$

$$\therefore T_E \approx 5.8 \text{ days}$$

The biological half-life of a given substance, reflecting its metabolic turnover, tends to vary from one individual to another and from one organ to another. It also depends on diet and disease. Accurate estimates of the biological (and hence effective) half-life are thus difficult to make and can cause serious problems regarding dosage assessments.

Emission of nuclear radiations

α-emission

An alpha particle (α-particle) is identical to a helium nucleus, or doubly-ionised helium atom (He^{++}), consisting of two protons and two neutrons. It thus carries a charge of $+2e$, and has a rest mass of about 7×10^{-27} kg.

Decay by α-particle emission occurs mainly amongst nuclei of the heavier elements of atomic number greater than that of lead ($Z_{Pb} = 82$) and results in a reduction of 2 in atomic number (Z) and 4 in mass number (A) of the radioisotope.

For example, radium decays to radon (Rn) in this way:

$$^{226}_{88}Ra \rightarrow ^{222}_{86}Rn + ^{4}_{2}\alpha$$

the emitted α-particle having a kinetic energy of about 4.8 MeV.

β^--emission

A negative beta particle (β^- or e^-) is simply a fast electron released from a nucleus during decay. It is termed a β particle mainly to distinguish it from orbital electrons. Radioisotopes having an excess of neutrons usually decay by this mode, during which a neutron is converted into a proton:

$$^{1}_{0}n \rightarrow ^{1}_{1}p + e^- + \bar{\nu}$$

neutron $\rightarrow$ proton + electron + antineutrino

The neutrino and its antiparticle, the antineutrino, are particles of zero charge and approximately zero mass which carry away a certain amount of energy and momentum from such disintegration processes.

During β^--decay, the radioisotope maintains its A but its Z increases by 1. For example, phosphorus decays to sulphur in this way:

$$_{16}^{32}P \rightarrow _{17}^{32}S + e^- + \bar{\nu}$$

Other radioisotopes commonly decaying this way include $_{17}^{35}S$, $_6^{14}C$ and $_{42}^{99}Mo$.

β^+-emission

A positive beta particle (β^+ or e^+) is called a positron, having the same rest mass as an electron but carrying an equal and opposite charge of $+e$. Neutron-deficient radioisotopes often decay by converting a proton into a neutron and emitting a positron:

$$_1^1p \rightarrow _0^1n + e^+ + \nu$$

$$\text{proton} \rightarrow \text{neutron} + \text{positron} + \text{neutrino}$$

In such a decay, A remains unchanged and Z decreases by 1. For example, sodium decays to neon as follows:

$$_{11}^{22}Na \rightarrow _{10}^{22}Ne + e^+ + \nu$$

$_9^{18}F$, $_{27}^{58}Co$, and $_{33}^{74}As$ also decay in this way.

Electron capture

A proton in the nucleus can capture an extra-nuclear electron from one of the inner electron shells, usually the K-shell, and combine with it to form a neutron:

$$_1^1p + e^- \rightarrow _0^1n + \nu$$

Such electron or K-capture, like β^+ decay, leaves A unchanged but decreases Z by 1. For example, chromium decays to vanadium in this manner:

$$_{24}^{51}Cr + e^- \rightarrow _{23}^{51}V + \nu$$

Other isotopes commonly decaying by electron capture include $_{26}^{55}Fe$ and $_{29}^{64}Cu$. Since an electron generally moves in from an outer orbit to fill the K-vacancy, a subsequent emission of characteristic X-rays accompanies K-capture.

γ-ray emission

γ-rays are emitted from a nucleus during transitions from an excited nuclear state to a lower-energy nuclear state. Such γ-ray emission often follows another decay process such as α- or β-emission, which has left the daughter nucleus in an excited state. Generally the γ-rays are emitted within a fraction of a microsecond of the primary decay, but sometimes there is a delay if the daughter nucleus is left in a metastable state. Such is the case with technetium-99m ($^{99m}_{43}$Tc), one of the most important radionuclides used in medicine. It is formed from molybdenum in a β^--decay process having a half-life of 67 hours:

$$^{99}_{42}\text{Mo} \rightarrow {}^{99m}_{43}\text{Tc} + e^- + \bar{v}$$

and subsequently decays by γ-emission with a half-life of 6 hours.

$$^{99m}_{43}\text{Tc} \rightarrow {}^{99}_{43}\text{Tc} + \gamma$$

Its short half-life demands that it is produced where it is to be used, and so a generator system or 'cow' is employed. This takes the form of an alumina column onto which is absorbed ^{99}Mo (the parent), which continuously decays to the daughter ^{99m}Tc. When required, the ^{99m}Tc is obtained by flushing the parent with a saline solution (elution), which washes out the daughter in high concentration without removing the parent. After the solution has been cleansed of surplus aluminium (from the column), ^{99}Mo and other impurities, it can be used either directly or in the formation of a labelled compound for tracer studies, (see Chapter 15).

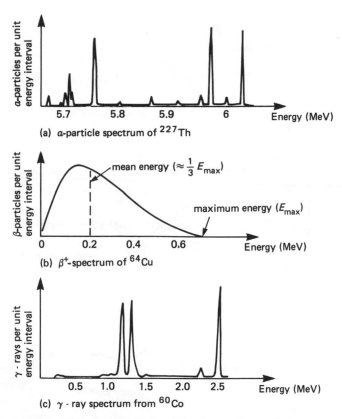

Figure 12.2   Energy spectra of the three nuclear radiations

Properties of nuclear radiations

Energy spectra

Both α- and γ-emissions occur at certain discrete energies, determined by the emitter and the particular decay route (see Fig. 12.2(a) and (c)).

In contrast, the β-particles emitted in radioactive decay have a continuous energy spectrum extending from zero to a well-defined maximum energy, which depends on the particular radioisotope concerned (Fig. 12.2(b)). This is because the energy available is shared between the emitted β-particle and its associated neutrino or antineutrino. Only when the latter's energy approaches zero does the β-particle acquire its maximum energy. β-energies can vary considerably from several keV in the decay of tritium to a few MeV from potassium decay.

Ionisation

Ionising radiation has the ability to remove orbital electrons from target atoms, thereby producing a number of ion pairs (ionised atom plus ejected electron) along its path. Directly-ionising particles are the charged particles (electrons, protons, α-

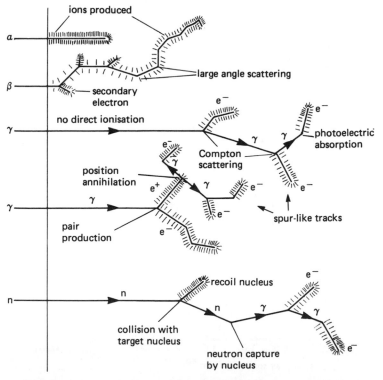

Figure 12.3 Ionisation tracks produced by nuclear radiations (diagrammatic)

particles, etc.) having enough energy to produce ionisation by collision. The uncharged particles (neutrons, photons) can liberate directly-ionising particles and are hence termed indirectly-ionising particles.

An average value for the ionisation energy of the constituents of air is about 34 eV. Since the energies of the nuclear radiations are far in excess of this, being commonly in the MeV region, the number of ion pairs produced along their paths in air is enormous.

(a) α-radiation

The α-particle is relatively slow-moving, has a high positive charge and large mass. It therefore easily tears electrons from target atoms leaving behind it a dense track of ionisation (Fig. 12.3). It suffers little deflection at each interaction and its path is thus characteristically straight.

(b) β-radiation

The β-particle is fast and light, and its initial, well-spaced collisions result in its deflection through large angles and the ejection of orbital electrons. The latter are known as secondary electrons and often have sufficient energy themselves to cause ionisation. As the primary electron gradually loses energy and slows down, the density of ionisation along its track increases until it is finally absorbed.

(c) γ-radiation

γ-radiation can produce ionisation in matter by three major mechanisms (see page 144):

  (i) photoelectric effect,
 (ii) Compton effect,
(iii) pair production,

the mechanisms predominating at low, medium and high photon energies respectively.

(d) Neutron radiation

In addition to the primary nuclear radiations, Fig. 12.3 also illustrates ionisation due to neutron radiation. Because of their lack of charge, neutrons can penetrate deep into target atoms and so interact with their nuclei. Fast neutrons (MeV range) are slowed down as a result of several such collisions with nuclei, which themselves recoil and produce tracks of dense ionisation. When the neutrons reach the eV range, they are termed slow or thermal neutrons, and are ultimately captured by target nuclei. The latter are normally unstable and γ-rays are emitted, in turn producing their own characteristic spur-like ionisation trails.

Penetration of matter

(a) α-radiation

Owing to the high density of ionisation it produces, the α-particle's penetration of matter is small. Further, since the α-particles from a given decay are monoenergetic and follow straight line paths, they all lose their KE completely after covering about the same distance, that is they have a definite range in a given medium (see

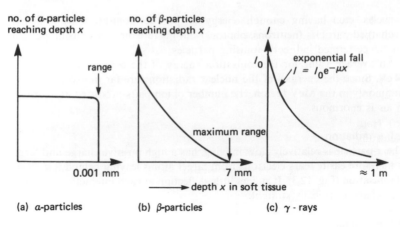

Figure 12.4   Penetration of 1 MeV nuclear radiation into soft tissue

Fig. 12.4(a)). In air, the ranges of 2, 5 and 10 MeV α-particles are approximately 10, 35 and 100 mm respectively; in aluminium 5–10 MeV α-particles penetrate about 0.01 mm.

Thus, α-particles from a source outside the body are relatively harmless, as they are stopped by the skin if not before. However, if α-particle emitters are ingested they can be extremely damaging, because of the dense ionisation they produce internally.

(b) β-radiation

β-particles are emitted with a variety of energies and do not follow straight line paths. Hence, their range in a medium is ill-defined, (see Fig. 12.4(b)). However, a maximum depth of penetration, or 'range', may be specified and clearly far exceeds α-penetration. For example 0.015, 3 and 5 MeV β-particles have maximum ranges of about 1 mm, 1 m, and 10 m respectively in air; whilst 1, 5, and 10 MeV β-particles have ranges of about 1.5, 10 and 20 mm in aluminium.

(c) γ-radiation

Despite their discrete energies at emission, γ-rays have no precise range in matter because they, like β-particles, follow tortuous paths. They are very penetrating and suffer an exponential fall in intensity as they traverse matter, (see Fig. 12.4(c)). If radiation of intensity I_0 falls on a medium, the intensity I after traversing a thickness x of the medium is

$$I = I_0 \, e^{-\mu x}$$

where μ is the linear attenuation coefficient of the medium (cf. equation [11.1]).

Although γ-rays therefore have no definite 'range' in a medium, it is useful to compare their penetration of matter with that of the other radiations by considering, for example, the thickness of absorber necessary to reduce the intensity of the γ-radiation to one tenth of its initial value. Then for 0.01, 1, and 10 MeV γ-rays, the thicknesses required are:

(a) 4, 300 and 1000 m respectively for air;

(b) 5, 300 and 1000 mm respectively for soft tissue;
(c) 0.4, 140 and 370 mm respectively for aluminium.

Similarly, 1 and 10 MeV γ-rays require:

(d) 30 and 40 mm of lead respectively;
(e) 150 and 410 mm of concrete respectively.

(d) Neutron radiation

Neutrons also have an approximately exponential transmission curve. They are very destructive of living tissue, being best absorbed in materials containing similarly 'sized' particles, such as the hydrogen atoms in water.

Production of artificial radioisotopes

Introduction

Hundreds of radioisotopes can be produced 'artificially' by a variety of methods, most of which strive to provide a product of as high a specific activity as possible. This allows the user greater flexibility since a radioisotope can always be 'diluted' by the addition of inactive carriers.

The methods largely involve the bombardment of certain stable nuclei by high-energy particles such as neutrons, protons, deuterons and α-particles, and the resulting nuclear reactions can lead to the formation of useful radioisotopes. Neutrons are particularly advantageous since their lack of charge enables them to penetrate target nuclei without experiencing any Coulomb repulsion.

Neutrons of sufficient energy to induce such nuclear reactions may be obtained in nuclear reactors and high-energy charged particles can be supplied from particle accelerators.

Nuclear reactor production methods

A fission reactor operating on natural uranium consists of a lattice of uranium rods embedded in a so-called moderating material. When the uranium fuel rods are irradiated with neutrons, the ^{235}U atoms capture neutrons to form the unstable ^{236}U, which then splits into two heavy fragments of approximately equal size during a process known as fission. For example:

$$^{235}_{92}U + ^{1}_{0}n \rightarrow ^{236}_{92}U \rightarrow ^{138}_{53}I + ^{95}_{39}Y + 3^{1}_{0}n + energy$$

A large number of useful radioisotopes are thus formed as fission products and can be chemically separated from the used uranium fuel rods of the reactor to yield relatively carrier-free radioisotopes of high specific activities. $^{90}_{38}Sr$ (strontium), $^{131}_{53}I$ (iodine), $^{137}_{55}Cs$ (caesium), $^{85}_{36}Kr$ (krypton) and $^{133}_{54}Xe$ (xenon) are amongst the numerous examples of such fission fragments. However, separation costs can be high due to the problems encountered through radiation hazard.

Much energy is released with each atom of uranium fissioned. A few fast neutrons are produced simultaneously with the fission fragments and these are then slowed down in the surrounding moderator, which is commonly carbon (graphite). The resulting slow neutrons are then captured by more ^{235}U nuclei and a chain reaction commences, with the release of enormous quantities of energy. In addition, vast numbers of neutrons are released, far in excess of those needed to sustain the chain reaction. The surplus can therefore be utilised to irradiate stable targets placed in the reactor and so produce radioisotopes. The transformation can occur in a variety of ways as described below.

(a) (n,γ) reactions

In the neutron-capture, or (n,γ) reaction, a target nucleus captures a neutron, and simultaneously emits a γ-ray, known as the capture gamma. The Z of the target atom remains unchanged but its A increases by 1. For example, for sodium the reaction proceeds as follows:

$$^{23}_{11}\text{Na} + ^{1}_{0}\text{n} \rightarrow ^{24}_{11}\text{Na} + \gamma$$

which may be abbreviated to:

$$^{23}_{11}\text{Na}(\text{n},\gamma)\ ^{24}_{11}\text{Na},$$

Other typical (n, γ) reactions include:

$$^{31}_{15}\text{P}(\text{n},\gamma)\ ^{32}_{15}\text{P}, \qquad ^{41}_{19}\text{K}(\text{n},\gamma)\ ^{42}_{19}\text{K},$$

$$^{58}_{26}\text{Fe}(\text{n},\gamma)\ ^{59}_{26}\text{Fe} \text{ and } ^{59}_{27}\text{Co}(\text{n},\gamma)\ ^{60}_{27}\text{Co}$$

Thus, the product radioisotope is chemically identical with the target, making separation by chemical means impossible. Since only a small fraction of the target atoms undergoes neutron capture, the radioisotope sample so produced contains a large percentage of carriers.

Sometimes however, the product itself subsequently decays, for example by β-emission, to yield a further radioisotope which is then chemically different from the target and from which it can be separated. For instance ^{131}I is produced from neutron-irradiated Te in the following way:

$$^{130}_{52}\text{Te}(\text{n},\gamma)\ ^{131}_{52}\text{Te}(\beta^{-})\ ^{131}_{53}\text{I}$$

(b) (n,p) reactions

During this reaction, in which the target nucleus gains a neutron and loses a proton, A remains unchanged but Z decreases by 1. The radioisotope ^{32}P of phosphorus is commonly produced from normal sulphur in the following way:

$$^{32}_{16}\text{S} + ^{1}_{0}\text{n} \rightarrow ^{32}_{15}\text{P} + ^{1}_{1}\text{p}$$

or briefly,

$$^{32}_{16}\text{S}(\text{n},\text{p})\ ^{32}_{15}\text{P}$$

Other examples include $^{14}_{7}\text{N}(\text{n},\text{p})\ ^{14}_{6}\text{C}$ and $^{35}_{17}\text{Cl}(\text{n},\text{p})\ ^{35}_{16}\text{S}$.

The radioactive product is now chemically different from the target and separation yields a nearly carrier-free radioisotope. Many radioisotopes required at high specific activity for clinical use are prepared in this way, the separation involving highly specialised techniques.

(c) (n,α) reactions

The (n,α) reaction involves the absorption of a neutron and emission of an α-particle, resulting in a reduction of 3 in A and 2 in Z of the target. Examples include:

$$^{35}_{17}Cl + ^{1}_{0}n \rightarrow ^{32}_{15}P + ^{4}_{2}\alpha$$

or more briefly,

$$^{35}_{17}Cl \,(n, \alpha)\, ^{32}_{15}P$$

Other examples include

$$^{27}_{13}Al(n, \alpha)\, ^{24}_{11}Na \text{ and } ^{6}_{3}Li \,(n, \alpha)\, ^{3}_{1}H$$

Rate of radioisotope production

The rate at which radioactive atoms are formed by bombarding a target with high-energy particles depends on a number of factors:

(a) Particle flux

The particle flux, ϕ, is the number of bombarding particles incident on the target per unit area per second. Obviously, the greater the flux, the greater the number of reactions.

(b) Number of target atoms

The number, n, of target atoms in a sample is given by:

$$n = \frac{mf}{A} \times 6.02 \times 10^{26} \qquad [12.6]$$

where m kg is the mass of the target element in the sample, A is the mass number of the target, 6.02×10^{26} kmol^{-1} is Avogadro's number, and f is the fractional isotopic abundance of the isotope involved in the reaction (the fraction of the target atoms which are the isotope in question).

Since the number of target atoms involved in reactions is small (less than one in a million) n remains almost constant during bombardment. Clearly, the number of reactions increases with n, which in turn depends on m, f, and A.

(c) Nuclear cross-section

Any nuclear reaction is characterised by a cross-section, σ, which describes the probability of the reaction occurring when the bombarding particle approaches the target nucleus. σ is expressed as the effective nuclear cross-sectional area through which an incident particle must pass for it to be 'caught' and for the reaction to occur. It depends on the target nucleus, the bombarding particle, and its energy. The greater the cross-sectional area, the more particles are 'caught' and more radioactive atoms result.

Thus, the rate X at which radioactive atoms are produced is given by:

$$X = \phi n \sigma \qquad [12.7]$$

Growth and decay of artificial radioisotopes

As soon as a radioisotope is produced, it starts to decay. Initially, the decay rate (and activity) are small, but these increase as irradiation continues and eventually saturation is approached when the production rate equals the decay rate:

$$X = \frac{dN}{dt} = a_{sat}$$

where a_{sat} is the saturation or maximum activity achievable. Using equation [12.7]:

$$a_{sat} = \phi n \sigma$$

and substituting from equation [12.6] gives:

$$a_{sat} = \phi \sigma \frac{mf}{A} \times 6.02 \times 10^{26}$$

Thus, the maximum specific activity obtainable is given by:

$$\frac{a_{sat}}{m} = \frac{\phi \sigma f}{A} \times 6.02 \times 10^{26}$$

Exercise 12

1 Explain the following terms:
  (a) nuclide,
  (b) stable isotope,
  (c) decay constant.
    The count rate from a radionuclide falls from 800 counts per minute to 100 counts per minute in 6 hours. What is the decay constant of the nuclide?

2 Define the activity and half-life of a radioisotope. What is the relationship between them?
    ^{131}I has a half-life of 8 days and ^{32}P has a half-life of 14.3 days. If carrier-free samples of these two radioisotopes initially have the same activity, sketch a graph to compare how their activities subsequently vary with time.
    The activity of 1 kg of ^{226}Ra is 3.7×10^{13} Bq, and its half-life is 1600 years. Find the activity of 1 kg of ^{131}I given that its half-life is 8 days. You may assume both radioisotopes are carrier-free. Why is the latter assumption necessary?

3 Distinguish between physical, biological and effective half-life, and state the relationship between them. What happens to the effective half-life when:
  (a) the physical half-life $\gg$ biological half-life;
  (b) the biological half-life $\gg$ physical half-life?
    Into which of these two categories, if either, do the following fall: ^{14}C, ^{24}Na, ^{3}H and ^{131}I?
    ^{14}C (carbon) is found in living organisms in the amount of about 100 atoms of ^{14}C for every 10^{20} atoms of ^{12}C. ^{14}C has a half-life of 5760 years. It is found that in a particular fossil, the amount of ^{14}C has decreased to about 10 atoms of ^{14}C for every 10^{20} atoms of ^{12}C. Estimate the age of the fossil. (Note: Carbon from carbon dioxide in the atmosphere is absorbed by living material only as long as it is alive.)

4 Describe the decay processes most likely to occur in:
(a) neutron-rich radioisotopes;
(b) neutron-deficient radioisotopes.
   Give one example of each.
   In the naturally occurring radioactive decay series, there are several examples in which a nucleus emits an α-particle, followed by two β-particles. Show that the final nucleus is an isotope of the original one. What is the change in mass number between the original and final nuclei?
   If the half-life of ^{226}Ra is 1600 years, calculate the number of years needed to reduce the activity of a given sample of ^{226}Ra to 3 per cent of its original value.

5 Discuss the possible emission processes taking place when a nuclide $^{A}_{Z}$X undergoes radioactive decay to a daughter nuclide:
(a) $^{A}_{Z-1}$Y
(b) $^{A}_{Z+1}$V
   A radon $^{222}_{86}$Rn nucleus of mass 3.6×10^{-25} kg decays by the emission of an α-particle of mass 6.7×10^{-27} kg and energy 5.5 MeV.
(c) What are the mass number (A) and atomic number (Z) of the resulting nuclide?
(d) What is the momentum of the α particle?
(e) Find the velocity of recoil of the resulting nucleus.
   (You may ignore relativistic effects.)

6 Explain the essential features of the energy spectra from typical α-, β- and γ-emitters. Would you expect α-, β- and γ-radiations of similar energy to be equally penetrating in matter? Give reasons.
   A source, having a half-life of 130 days initially contains 10^{20} radioactive atoms and the energy released per disintegration is 8×10^{-13} J. Calculate:
(a) the activity of the source after 260 days;
(b) the total energy released during this period.

7 Explain the role of the nuclear reactor in the production of radioisotopes. Describe three nuclear reactions which are used to produce radioisotopes in a reactor, and give a clear example of a radioisotope produced by each method.
   What is meant by a carrier-free radioisotope? Discuss the relative success with which the methods you have described achieve such carrier-free radioisotopes.
   A small volume of a solution which contained the sodium radioisotope ^{24}Na had an activity of 12 000 disintegrations per minute when it was injected into the blood stream of a patient. After 30 hours, the activity of 1 cm^3 of the blood was found to be 0.5 disintegrations per minute. If ^{24}Na has a half-life of 15 hours, estimate the volume of blood in the patient.

8 Describe in detail the factors which govern the amount of a radioisotope produced in a nuclear reactor.
   Two radioactive sources X and Y initially contain the same number of radioactive atoms. Source X has a half-life of 15 minutes and source Y a half-life of 30 minutes. What is the ratio of the activity of X to that of Y;
(a) initially;
(b) after 30 minutes;
(c) after 2 hours?

13 | Radiation units and dosimetry

Biological effects of ionising radiation

The way in which ionising radiation interacts with biological materials is not yet fully understood. Ionisation is thought to lead to the alteration of essential cell molecules such as enzymes and DNA, so disturbing their function. At the same time, new constituents formed may exert catalytic or toxic effects on metabolic processes.

The sensitivity of cells to radiation damage depends on the type of cell. For example, cells of the reproductive organs and intestines are radiation-sensitive, whilst bone and nerve cells are relatively radiation–resistant. The major injury seems to be to the reproductive mechanism of the cell, resulting in mutation, sterilisation or even outright destruction, and explains the delayed appearance of damage so typical of radiation effects.

Gross body effects resulting from radiation exposure range from the overt diseases suffered by early workers, including skin burns, radiation sickness (diarrhoea, nausea, vomiting, inflammation of the throat, loss of hair, loss of appetite, fever, pallor, rapid emaciation and eventual death), warts and deformed fingers, to the more latent conditions such as anaemia, leukemia, cancer and sterility.

Radiation exposure

When X- or γ-radiation traverses matter it is found that a measurement of the ionisation produced provides a good indication of the total energy absorbed. Since the energy absorbed per photon per kilogram depends only on photon energy, E, and the atomic number of the medium, the energy absorbed per kilogram in air (effective $Z = 7.6$) closely resembles that in living tissue (Z for muscle $= 7.4$).

Thus, ionisation in air has long been accepted as a basis for X- and γ-ray monitoring, and the term exposure is accordingly defined: if Q is the total charge of the ions of one sign produced in air when all the β-particles (electrons and positrons) liberated by photons in a volume of air of mass m are completely stopped in air, then:

$$\text{Exposure, } X = \frac{Q}{m}$$

The unit of exposure is therefore $C\,kg^{-1}$.

Despite its simplicity, the term exposure has limited applications, since:

(a) it applies only to X- and γ-radiation;
(b) strictly speaking it refers to *ionisation* in *air* and not to *energy absorption* in *living materials*.

Absorbed dose

If E is the mean energy imparted by ionising radiation to a volume of material of mass m, then:

$$\text{Absorbed dose, } D = \frac{E}{m}.$$

The unit of absorbed dose is thus $J\,kg^{-1}$, otherwise known as the gray (Gy). This unit describes all kinds of radiation being absorbed in all types of material, and is thus a more versatile unit than the $C\,kg^{-1}$. However, direct measurement of absorbed dose is difficult and it is usually calculated from a measurement of exposure.

Relationship between exposure and absorbed dose

When X- or γ-radiation traverses air, an exposure X of $1\,C\,kg^{-1}$ produces $1/e$ electrons per kg of air, where e is the electronic charge. Since the average ionisation energy for air is about $34\,eV$, ($= (34 \times e)\,J$), an exposure of $1\,C\,kg^{-1}$ results in an approximate energy absorption of:

$$\frac{34\,e}{e} = 34\,J\,kg^{-1}, \text{ or } 34\,Gy.$$

Therefore, for air:

$$D(Gy) \approx 34 \times X(C\,kg^{-1})$$

The energy absorbed per kilogram of any other material depends on its particular absorption properties. Using $f(JC^{-1})$ as a general conversion factor:

$$D(Gy) = f \times X(C\,kg^{-1})$$

Figure 13.1 illustrates how f varies with photon energy for bone, muscle, water (approximating soft tissue) and air. Such conversion aids are useful, since in practice air ionisation chambers, measuring X in $C\,kg^{-1}$, are often used to monitor radiation dosages (D). At high photon energies, where the Z-independent Compton scatter dominates, f assumes about the same steady value for all body materials. However, at lower photon energies the highly Z-dependent photoelectric effect dominates, and f varies significantly from one medium to another. In this energy range, identical *exposures* lead to widely different *absorbed doses* in different media, making accurate dose assessments difficult.

Exposure rate and absorbed dose rate

$$\text{Exposure rate} = \frac{dX}{dt}\,C\,kg^{-1}\,s^{-1}$$

where dX is the increment in exposure in the time interval dt. This unit is used in ionisation instruments which register electric current (rate of collection of charge)

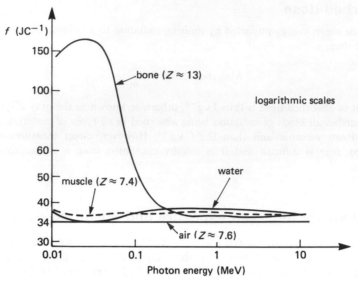

Figure 13.1   Relationship between exposure and absorbed dose

rather than simply charge, which gives exposure in $C\,kg^{-1}$. Similarly:

$$\text{Absorbed dose rate} = \frac{dD}{dt} \text{ Gy s}^{-1}$$

where dD is the increment in absorbed dose in the time interval dt.

Dose equivalent

Biological damage is not solely dependent on absorbed dose. Heavily ionising radiations, such as α-radiation, can cause more cellular damage than the less densely ionising radiations like β- or γ-radiation, even when the energy absorbed per kilogram (absorbed dose) is the same in each case.

The effectiveness of a radiation in producing biological damage is measured in terms of a dimensionless quality factor, Q. Table 13.1 lists, for the common radiations, the values of Q recommended by the International Commission on Radiological Protection, ICRP.

Table 13.1 Quality factors

Type of radiation	Q
X-, γ- and β-radiation	1–2
Slow neutrons	5
Fast neutrons, protons and α-particles	10
Heavy recoil nuclei	20

The biological effectiveness of a radiation also depends on the irradiation conditions, in particular the distribution of the absorbed dose in space and time. If all such remaining modifying factors are grouped together. in another dimensionless quantity N, the dose equivalent of the radiation H is defined as:

$$H = \text{absorbed dose} \times Q \times N$$

Dose equivalent thus describes the relative radiation risk resulting from a particular radiation. Like absorbed dose, its basic unit is $J\,kg^{-1}$, but to distinguish the two, dose equivalent is assigned the unit known as the sievert (Sv).

$$H(Sv) = D(Gy) \times Q \times N.$$

At present, N is taken as 1 for external sources, but its value may be higher for ingested radioactive material.

Radiation levels

Table 13.2 Common irradiation situations

Source	Dose equivalent (mSv)
Natural external background radiation (UK)	1 (per year)
Internal exposure to ^{226}Ra and ^{40}K from foods	1–5 (per year)
One chest X-ray:	
best	0.1
average	2
fluoroscopic examination	100
Local dose during therapy	30 000–70 000

There are many sources of background radiation, including cosmic rays and naturally-occurring radioisotopes, both in the body (^{14}C and ^{40}K) and in rocks, and buildings (^{238}U and ^{226}Ra). Sharp increases in background radiation can occur, for example, after nuclear testing, and depend very much on location.

The ICRP has recommended maximum permitted dose levels (MPLs) for various situations. These depend on a number of factors, including the age, sex, condition, and occupation of the irradiated individual, as well as on the type of radiation and its region of application. For example, the MPL for a radiation worker is 50 mSv per year (whole body or reproductive organs), rising to 750 mSv per year (hands, forearms, feet). The non-radiation worker has an MPL of only 5 mSv per year (whole body), and even stricter limits are set for pregnant women and children.

Old units and conversions

Table 13.3 Unit conversions

Quantity	New unit	Old unit	Conversion
Activity (source)	Bq	curie (Ci)	$1\,Bq \approx 2.7 \times 10^{-11}\,Ci$
Exposure	$C\,kg^{-1}$	röntgen (R)	$1\,C\,kg^{-1} \approx 3876\,R$
Absorbed dose	Gy	rad	$1\,Gy = 100\,rad$
Dose equivalent	Sv	rem	$1\,Sv = 100\,rem$

Exercise 13

1 Distinguish between exposure and absorbed dose, and define the units in which they are measured.

Discuss the factors which determine what absorbed dose is received for a given exposure.

The activity of a carrier-free sample of ^{32}P is found to be $3.7 \times 10^{10}\,Bq$. If its half-life is 14.3 days, calculate the number of radioactive atoms present.

2 Define the following units:
(a) gray (Gy),
(b) sievert (Sv),
(c) becquerel (Bq).

An experiment is carried out to investigate the effect on tissue of a neutron beam and a γ-ray beam, having quality factors of 10 and 1 respectively. It is possible to insert a lead shield into the γ-ray beam for which the half-value thickness in lead is 10 mm. With the shield in position, it is found that both beams produce the same absorbed dose. Without the shield, the two beams produce the same dose equivalent in the tissue sample. What is the thickness of the lead shield?

3 Give some common sources of irradiation of the human body and indicate their relative importance.

What are some of the effects which can result from excessive radiation exposure? Discuss possible means of protection against such radiation hazards.

A carrier-free sample of ^{131}I has a mass of $10^{-3}\,kg$. If its half-life is 8 days, find its activity in Bq. (Avogadro's number is 6.02×10^{26} atoms per k mol.)

4 (a) Explain the meaning of the terms in italics in the following sentence:

A radioactive source containing material whose *activity* is $3 \times 10^9\,Bq$ ($8 \times 10^{-2}\,Ci$), emits only γ-rays whose energy is such that the *half-value thickness* in lead is $9 \times 10^{-3}\,m$ and produces at 1 m from the source an *absorbed dose* in tissue of $5 \times 10^{-3}\,Gy$ (0.5 rad) every hour.

(b) A source such as that described in (a) falls from its protective container onto the floor. It has to be replaced by a technician who has available a pair of long-handled tongs and a lead shield 18 mm thick, behind which he can work. Tests made previously have shown that it takes 36 s to replace the source if the lead shield is used, but only 8 s if no shield is used. In each case the body is the same distance from the source. Neglecting other factors which might affect the problem, state, explaining your reasons, whether or not you would recommend the technician to use the lead shield.

(c) Suppose another, longer, pair of tongs became available, enabling the body to be kept twice the distance away from the source that was possible in (b), but with which no shield can be used. State, with reasons, whether you would recommend the use of the longer tongs if it takes 27 s to complete the operation. [JMB]

5 (a) Define the unit of radiation exposure. Describe the mode of action of an ionization chamber and explain how it is used to measure exposure.

(b) Define the unit of absorbed dose and calculate the absorbed dose in air when the exposure is one unit.

(c) A neutron beam and a beam of gamma rays each produce the same absorbed dose in a certain body. State whether the dose equivalents are the same and give a reason for your answer.

The charge of an electron $= -1.6 \times 10^{-19}$ C

Energy required to produce one ion-pair in air $= 34$ eV

[JMB]

6 (a) Define the unit of *absorbed dose*.

Draw a typical spectrum of the X-radiation produced from the target of a diagnostic X-ray tube.

Indicate on your diagram that part of the spectrum usually removed by filtration. Explain the reason why filters are used and suggest a suitable material for such a filter.

(b) List **three** factors which affect the quality of a radiographic image produced by X-rays. Explain how, instead of allowing the radiation to form an image directly on suitable film, it is possible to reduce the absorbed dose to the patient and still obtain a photographic image of the same density.

[JMB]

7 (a) The effects of ionizing radiations on a biological sample depend not only on the absorbed dose but also on the types of ionizing radiations employed.

Explain the terms *relative biological effectiveness* and *dose equivalent*.

State **two** effects which may be observed in human beings and *two* which may be observed in individual cells exposed to ionizing radiations.

Considering alpha particles and electrons, state

(i) which is more damaging to tissue for equal absorbed doses,

(ii) an important difference in their behaviour which is related to the amount of damage caused.

(b) State, without explanation, *three* methods which are used for minimising the absorbed dose to patients during diagnostic radiography.

Calculate the ratio of the transmitted to the incident intensities of an X-ray beam travelling through a layer of aluminium 2 mm thick, the half-value thickness being 3 mm. [JMB]

14 | Radiation detectors

Ionisation chambers

General principles

Ionisation chambers are instruments designed to detect and measure radiations through their ionising effect on matter. The chamber, which is gas-filled, often with air, contains two electrodes, and may, for example, be a cylindrical type, comprising a cylindrical cathode with a central wire anode, or alternatively a parallel-plate type.

When ionising radiation enters the chamber, ion pairs are produced and these are attracted to the oppositely-charged electrodes. If the voltage V across the electrodes is low, many of the ions produced by irradiation recombine before they reach the electrodes, and the resulting ionisation current is small. As V is increased, however, less and less recombination occurs until eventually all the ions produced are collected and a 'saturation' current is obtained. The chamber is then said to be operating in the 'ionisation region'.

The ionisation current is normally very small $(10^{-10}–10^{-15}\,\text{A})$ and sensitive instruments are demanded for its detection. The ionisation chamber is unsuitable for measuring low exposure rates since an inordinately large chamber would be necessary to yield a sufficient ionisation current.

If the chamber is not hermetically sealed, accurate readings need atmospheric pressure and temperature corrections. For example, if the calibration is correct at 760 mmHg and 293 K, a pressure of P mmHg and T K leads to a corrected reading, given by:

$$\text{corrected reading} = \text{meter reading } \times \frac{760}{P} \times \frac{T}{293}$$

The major advantage of the ionisation chamber is its great accuracy (± 0.1 per cent under favourable conditions) which is not equalled by any other radiation detector. Other useful features include a good energy response (i.e. output little effected by the energy of the incident radiation) and versatility, as illustrated in Table 14.1 which lists some common designs of ionisation chambers together with their particular applications. Details of each type are given below.

Free-air ionisation chamber

An instrument to measure exposure, Q/m, should by definition collect all the charges of one sign which the secondary electrons, generated in a known mass or volume of air, can produce in air before being stopped. The free-air ionisation

Table 14.1 Ionisation chambers

Instrument	Application
Free-air chamber	Primary standard of exposure
Thimble chamber	Exposure and absorbed dose measurements
Capacitance thimble chamber	Personnel radiation monitoring
Monitor chamber	Checking of machine output
Pocket ionisation chamber	Personnel radiation monitoring
Neutron ionisation chamber	Measurement of neutron irradiation

chamber (Fig. 14.1) is a standard instrument designed for this purpose. It consists of a large lead-lined chamber containing two parallel-plate electrodes insulated from each other, and across which a potential difference is maintained by a high tension supply. The ionising radiation enters through an aperture of precisely known dimensions in a diaphragm on one side of the chamber and the resulting narrow collimated beam ionises air in its path. The ions produced between the plates are collected at the two electrodes and the accumulated charge is recorded on a sensitive electrometer (e.g. quartz-fibre, valve or vibrating reed) connected to one of the plates.

The lines of force between the plates are parallel in the middle, but bowed at the edges, making an accurate assessment of the volume of air from which ions are collected difficult. Therefore, one of the plates is split into three, with very narrow gaps between, and the outer sections or guard plates are earthed and not connected to the electrometer. Thus, ions are only collected from the well-defined

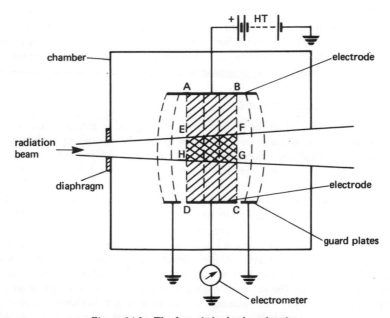

Figure 14.1   The free-air ionisation chamber

volume ABCD, corresponding to the initial ionised air volume EFGH. The latter volume is used to calculate the mass (m) of air to insert in the expression:

$$\text{exposure} = \frac{Q}{m}$$

where Q is the charge registered by the electrometer.

A number of points should be noted:

(a) Some secondary electrons starting within the 'collection region' ABCD leave it and ions produced along their paths are not collected. On the other hand, some electrons generated outside the region EFGH enter the region ABCD and produce ionisation there. It is assumed that there is electronic equilibrium, i.e. as many secondary electrons leave as enter the collection region.

(b) All the ions must be collected before they have a chance to recombine, i.e. saturation must occur. This is achieved by increasing the voltage across the plate electrodes to an appropriate value which depends on the exposure rate and the size of the chamber. Potential differences between 10 and 20 volts per millimetre of plate separation are required for exposure rates below about $1 \, \text{C kg}^{-1}$ per hour.

(c) The spacing of the plates must be sufficient to allow all the secondary electrons produced within EFGH to complete their paths in air rendering them incapable of producing further ionisation before arriving at either plate. The higher the radiation energy, the higher the maximum energy of secondary electrons, and hence the greater must be the plate separation. For example, for $100 \, \text{kV}_p$ X-radiation, separations of about $0.25 \, \text{m}$ are adequate, whilst γ-radiation of around $2 \, \text{MeV}$ requires a spacing nearer $2 \, \text{m}$, unless the air pressure is raised.

It is clear that the free-air chamber is too large for routine medical use but its great accuracy makes it suitable for use as a primary standard in standards laboratories like the National Physical Laboratory. Other secondary standards are then calibrated against such an instrument.

Thimble ionisation chamber

The thimble ionisation chamber is an instrument designed to monitor exposure, or exposure rate, at a point on or in a patient or 'phantom' simulating the patient. It is normally very small (of volume less than $1 \, \text{cm}^3$) and accepts radiation from a wide range of directions. A larger model (of volume about 1 litre) is sometimes used to survey equipment and rooms for stray radiation.

Figure 14.2 illustrates a typical thimble chamber. A small known volume of air is enclosed in the chamber, the walls of which are solid but made of a material exhibiting radiation absorption properties similar to those of air if compressed to the same density. Such a substance is known as an air-equivalent material, and has an effective atomic number and electron density approximating to those of air. Graphite ($Z = 6$) has been widely used in the past, but a bakelite–graphite mixture is now more common. The resulting chamber is then known as an 'air-walled' or 'air-equivalent' chamber. The thickness of the wall is just sufficient to stop the passage of all secondary particles from outside, but not so great that excessive

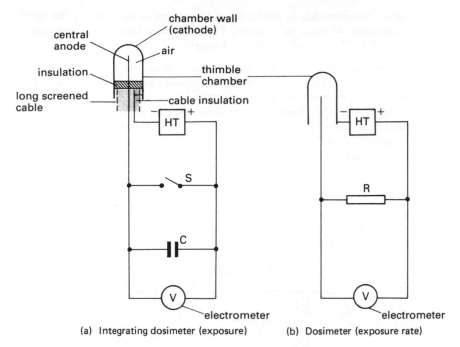

Figure 14.2 Thimble chamber circuits

absorption in the wall reduces the ionisation inside the chamber. Thus, the ionisation in the small air volume is due to secondary particles generated by radiation absorption in the wall and to a small extent in the air itself.

The surface of the wall is made electrically conducting so that it forms one electrode of the chamber. The other electrode is a central thin rod, often of aluminium, which is insulated from the chamber wall by material such as amber.

A cable connects the electrodes to an external circuit, which supplies a voltage across the electrodes sufficient to ensure saturation, and measures the resulting ionisation in one of two ways:

(a) The ionisation charge Q liberated during irradiation of the chamber causes the capacitor C to charge to a voltage V, which is recorded by the electrometer. Since:

$$V = \frac{Q}{C}$$

the electrometer reading is proportional to Q, which continues to rise as long as radiation is received. The instrument, termed an integrating dosimeter, thus measures total exposure in $C\,kg^{-1}$. The switch S is included to discharge C between readings.

(b) An electrometer measures the voltage V developed across the very high resistor R ($10^{10}-10^{13}\,\Omega$) by the ionisation current i. Since $V = iR$, the electrometer reading is proportional to i, which in turn reflects the rate of production of ionisation charge. The instrument thus measures exposure rate $(C\,kg^{-1}\,s^{-1})$ and is called a dosimeter.

The electrometer is calibrated to read exposure or exposure rate directly either by comparing the readings with those from a free-air chamber, or by exposing the thimble chamber to standard radiation sources.

Although the thimble chamber shows some inaccuracies at very high and very low energies, its useful range may be extended using a thinner-walled chamber for low energies and a 'build-up cap' (e.g. 2–4 mm of perspex) over the chamber for the higher-energy radiations such as ^{60}Co γ-radiation.

Capacitor (or condenser) ionisation chamber

In the cable-less capacitor ionisation chamber (Fig. 14.3) removal of a screw in the outer case allows electrical contact to be made with the central electrode. Initially, the chamber is charged to a voltage V_1 measured by an electrometer. The screw is

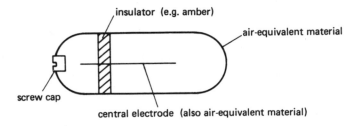

Figure 14.3 Capacitor ionisation chamber

replaced and the chamber exposed to radiation. The resulting ionisation leads to a gradual discharging of the chamber, so that when the screw is removed, the new voltage V_2 recorded is less than V_1. If C is the capacitance of the chamber:

$$\text{loss of charge} = C\,(V_1 - V_2)$$

This is equivalent to the total ionisation charge, and hence:

$$\text{exposure} \propto (V_1 - V_2)$$

The electrometer scale can be calibrated to register exposure $(C\,kg^{-1})$ directly. The sensitivity of the instrument depends largely on its size, a larger chamber giving greater sensitivity. It is important to ensure that the chamber is never completely discharged during exposure, since the p.d. must be sufficient to maintain saturation.

The great advantage of the instrument is the speed and simplicity with which it allows exposure to be assessed. On the other hand, it is fragile and relatively costly.

Monitor ionisation chambers

The ionisation chambers employed to monitor the output from radiation machines are preferably of the parallel-plate type. The axis of the beam then strikes the plates perpendicularly, ensuring uniform beam filtration, and avoiding any non-uniform shadows which might be cast using instruments like the thimble chamber. Suitable calibration against a primary (free-air chamber) or secondary (thimble chamber) standard is of course necessary.

Pocket ionisation chamber

A popular personnel radiation monitoring device is the pocket ionisation chamber (PIC) (Fig. 14.4) which consists basically of three parts:

(a) a capacitor ionisation chamber;
(b) a built-in quartz-fibre electrometer;
(c) a microscope.

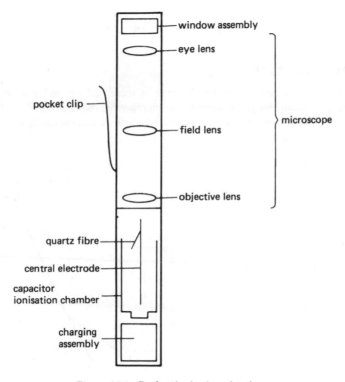

Figure 14.4   Pocket ionisation chamber

The instrument is first plugged into a charging unit via its charging assembly and the capacitor ionisation chamber acquires its requisite charge. Attached to its central electrode is a quartz-fibre in the form of a hairpin movement and when the chamber is fully charged the fibre points to zero on its associated scale. When the chamber is irradiated, it begins to discharge and the fibre moves across the scale calibrated directly in $C\,kg^{-1}$ or Gy. The scale which is viewed directly through the incorporated microscope system, generally has a range of about $0-2 \times 10^{-3}\,Gy$.

The advantages of the PIC are that it:

(a) gives an accurate, direct reading of exposure;
(b) is small enough to be clipped onto a pocket;
(c) is independent of radiation energy over quite a large range.

On the other hand it is relatively costly, provides no permanent record of exposure and gives no information about the type of radiation received.

Neutron ionisation chambers

The standard ionisation chamber can be made sensitive to neutron irradiation by lining the inner wall of the chamber with a material such as boron. The boron atoms capture the neutrons and liberate α-particles, which then ionise the air in the chamber.

$$^{10}_{5}B + ^{1}_{0}n \rightarrow ^{7}_{3}Li + ^{4}_{2}\alpha \text{ or } ^{10}_{5}B(n,\alpha) \, ^{7}_{3}Li$$

Radiation counters

If the voltage across an ionisation chamber is increased beyond the saturation region, the ionisation charge or current also increases as indicated in Fig. 14.5. This is due to the primary ions being so accelerated towards their respective electrodes

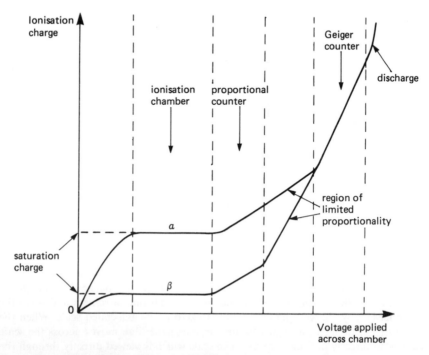

Figure 14.5   The variation of collected charge across ionisation detectors

that they can produce secondary ionisation of the gas molecules by collision. This 'avalanche' process produces a multiplication of the original ionisation by factors as high as 10^3 or 10^4, so that the resulting charge (or current) is much greater than the saturation value. Detectors operating at and above these voltages, when used in conjunction with suitable amplifiers and sensitive measuring instruments, can register the arrival of single ionising particles. It is then more usual to describe the output from such detectors in terms of a pulse size or height and to refer to the instrument as a counter.

Immediately beyond the saturation region, there is a region in which the pulse size is proportional to the energy lost by the original particle. In other words, in this proportional region, the original ionisation is multiplied by a constant factor, and a detector operating in this region is called a proportional counter.

If the applied voltage is increased beyond the proportional region, eventually a stage is reached where all pulses are magnified to a constant size regardless of the initial amount of ionisation caused by the ionising particle or photon. Thus, the two curves for α-radiation (heavy initial ionisation) and β-radiation (light initial ionisation) come together in the so-called Geiger region. An instrument operating in this region is called a Geiger counter.

Geiger counters

Structure and operation

The Geiger (or Geiger–Müller) counter (Fig. 14.6) consists of a tubular envelope of metal or glass containing a suitable gas, usually argon, at a low pressure of about 100 mmHg. The central anode is a thin wire (e.g. tungsten) and the cathode may be either in the form of a coating of conducting material such as graphite or silver deposited on the envelope, or alternatively a metal cylinder supported inside the envelope.

The cathode is earthed and the anode is held at a high positive potential

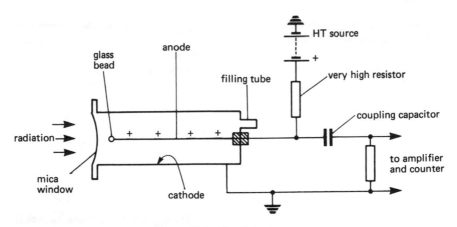

Figure 14.6 The Geiger counter

($\approx + 1000$ V). The glass bead at the end of the anode prevents sparking from points at the end of the wire.

If an ion pair is produced in a Geiger counter, the electron moves towards the central anode and the positive (argon) ion moves more slowly towards the cathode. An avalanche of electrons builds up through secondary collisions, and because the electric field near the wire is so intense, the discharge spreads along the whole anode. This renders the charge available for collection by the wire of constant value, independent of the magnitude and location of the initial ionisation. A short-duration pulse of constant size therefore results from the discharge.

Quenching

During the discharge at the anode, the positive ions are gradually approaching the cathode and if they reach it, they have sufficient energy to eject electrons from it. These would initiate further discharges which would lead to spurious pulses. This action is prevented by quenching, which may be achieved by either chemical or electronic means.

(a) Chemical
Alcohol, chlorine or bromine vapour is added to the argon as a quenching agent. Alcohol, for example, has a lower ionisation potential than argon, and the ionised argon atoms therefore acquire electrons from the molecules of alcohol, which then arrive at the cathode as positive ions. The energy available as they become neutralised at the cathode is then absorbed in dissociating the alcohol molecules rather than in ejecting electrons from the cathode.

(b) Electronic
An external circuit can lower the anode voltage below the limit required to sustain the discharge for a few hundred microseconds after the arrival of a pulse.

Errors

Following the entry of an ionising particle and its resultant pulse, the counter has a fairly long insensitive, paralysis or resolving time of about 0.1 ms. This is made up of a dead time (no counts recorded) and a recovery time (pulses of reduced size recorded). However, corrections can be made to the recorded count to allow for the undetected particles that enter the counter while it is inoperative.

The great sensitivity of Geiger counters, making them suitable for detecting very low activities, also makes them susceptible to errors from, for example, background radiation. The background count, due to radioactive contaminants in the detector, cosmic radiation and 'environmental radiations', can be minimised by operating the counter inside a lead shield.

Counter efficiency

For α- and β-radiation, the Geiger counter's efficiency for those particles which reach the sensitive volume is approximately 100 per cent. The problem is preventing excessive attenuation of the beams in the walls or window of the counter, and for this reason counters with very thin mica windows are used to allow weak β-radiation and possibly even α-radiation to penetrate and be counted.

The detection of γ-rays, however, depends on the conversion of the radiation into electrons in the counter walls since the production of electrons in the low-density gas filling is very small. Thus, for γ-radiation the efficiency depends on the atomic number and thickness of the counter wall (commonly of lead) and is generally only about 1 per cent for photon energies around 1 MeV. Some improvements can be obtained by making the cathode of a high-Z material, such as copper or lead, and increasing its area using an attached mesh so that more secondary electrons are liberated into the gas from the cathode.

Uses and relative merits

The most common uses of the Geiger counter in medicine include:

(a) Measurement of β-radiation, including that inside the body using a special miniature counter, about 20 mm long and 2 mm in diameter. This is particularly useful for monitoring the radiation from pure β-emitters like ^{32}P introduced into the body.

(b) Monitoring of equipment to detect contamination. A pocket-sized counter incorporating a buzzer may be employed as a warning device.

(c) Counting of charged particles in the presence of γ-radiation, (since the sensitivity to the latter is so low).

The counter is stable, reliable, very sensitive (and can therefore record low activities) and provides an output pulse of several volts which needs no further amplification. However, it suffers from a long paralysis time, gives no information about particle type, and quenching agents periodically need replacing.

Scintillation counters

Basic principles

Ionising radiation may be detected by the scintillations it produces in certain materials known as scintillators, or phosphors (see page 106). Since the intensity and duration of an individual scintillation are too small for convenient observation and counting, the effect is magnified using a photomultiplier tube. The combination of scintillator and photomultiplier is then known as a scintillation counter (Fig. 14.7).

The initial amount of light energy generated is approximately proportional to the energy deposited in the phosphor. Some of these light photons then fall on the light-sensitive cathode, or photocathode, of the photomultiplier tube and by the photoelectric effect eject electrons from its surface. This group of photoelectrons constitutes the primary electrical signal, which is then amplified by a factor of between 10^5 and 10^8 in the photomultiplier tube. Since each initial pulse of light is amplified by the same factor, the output electrical pulse from the photomultiplier is approximately proportional to the original energy deposited in the phosphor. By employing a further amplifier and a multi-channel pulse-height analyser in the output circuit, it is possible to count the number of pulses in a given energy range and so obtain an energy spectrum corresponding to that of the incident radiation.

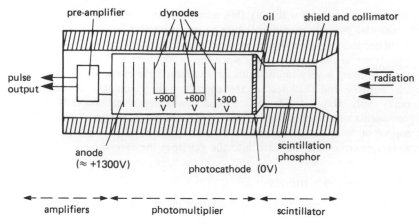

Figure 14.7   The scintillation counter

The scintillator

A number of solid or liquid materials may be used as scintillator and they may be broadly divided into three classes:

(a) inorganic scintillators: ionic crystals such as zinc sulphide (to detect α-particles), sodium iodide (for γ-radiation), and lithium iodide (for thermal neutrons);

(b) organic scintillators: large molecular crystals such as anthracene, used mainly for β-particles and fast neutrons;

(c) ·solution scintillators: luminescent materials in
   (i) solid solution (plastic scintillators, useful for high β-count-rates),
   (ii) liquid solution (liquid scintillators, particularly useful for monitoring low-energy β-emitters, like ^{35}S, ^{14}C and ^{3}H. The direct mixing of the sample with the scintillator permits a high counting efficiency without the loss of counts through self-absorption of radiation in the sample.)

Of all the scintillators, perhaps the most widely used is sodium iodide, activated with about 0.5 per cent of thallium iodide, the abbreviation being NaI (Tl). Since NaI is hygroscopic, the crystals are kept hermetically sealed, and are usually purchased in sealed aluminium cans, having a glass or quartz window on the side to be in contact with the photomultiplier. Good optical contact is essential between the window and face of the photomultiplier and such optical 'coupling' is usually achieved using silicone oil or clear grease.

NaI has many advantages, including:

(a) a relatively high efficiency (≈ 10 per cent) for converting particle energy into light; this permits detection of low-energy radiation and favours good energy resolution;

(b) a short scintillation duration enabling high count-rates (up to about 10^4 counts per second) to be recorded;

(c) excellent γ-ray detection, due to its high density (which increases the probability that an incident γ-photon will interact with an atom in the crystal and produce a scintillation) and its relatively high atomic number ($Z = 53$ for iodine, which favours photoelectric absorption and the generation of a pulse representing the full photon energy).

A common arrangement for measuring the activity of liquid samples utilises a well-type NaI (Tl) scintillator detector head (Fig. 14.8).

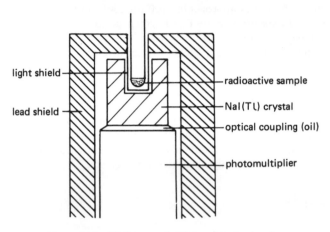

light shield

lead shield

radioactive sample

NaI(Tl) crystal

optical coupling (oil)

photomultiplier

Figure 14.8   Well-type scintillator detector head

The photomultiplier

Light photons from the scintillator fall on the photomultiplier's photocathode, a thin layer of semiconductor material (usually antimony-caesium) deposited on the inner surface of the photomultiplier tube face. A certain percentage, (≈ 10–25 per cent), of the incident photons eject photoelectrons from the photocathode, which is normally earthed. These photoelectrons are then accelerated towards the first of a series of electrodes or dynodes which are held at increasing positive potentials along the tube (Fig. 14.7). Each dynode is made from low work function material such as a beryllium–copper alloy.[1] Thus, an energetic electron incident on its surface easily ejects from it a few secondary electrons. These in turn are accelerated to the next dynode where further secondary electrons are ejected, and so on down the chain of normally ten dynodes until the final anode is reached. If there is an average multiplication of four at each of the ten dynodes, every primary photoelectron will result in a pulse of 4^{10} ($\approx 10^6$) secondary electrons at the anode. Such a large overall gain enables even low-energy radiations to be detected and measured.

A finite background count (dark current) is always present due to thermionic emission from the photocathode and dynodes, as well as stray radiations. The latter are largely prevented by shielding, and the former may be reduced by cooling the photomultiplier.

---

[1](The work function of a metal is the minimum amount of energy which must be supplied to an electron to liberate it from the surface.)

Principle uses of the scintillation counter

The scintillation counter is a very versatile instrument, capable of measuring α-, β-, γ-, X- and neutron-radiation by the selection of a suitable phosphor of appropriate shape and size. Common applications include:

(a) X- and γ-radiation detection, for which it is the preferred instrument: counting efficiencies approaching 100 per cent can be achieved;
(b) analysis of γ-spectra, (e.g. see Fig. 12.2(c) page 167), often with the intention of identifying the source;
(c) low intensity monitoring, typical of tracer investigations (see gamma camera page 208);
(d) liquid sample counting, using either a liquid scintillator or a well-type scintillator crystal.

Disadvantages include relatively high cost and size. The scintillator diameter varies from about 1–10 cm depending on the photon energies to be recorded, and cannot be further reduced without adversely affecting accuracy, since a photon energy may not then be totally absorbed in the crystal.

Film badges

Structure

The film badge dosimeter is the most common method of personnel monitoring and uses the blackening of a photographic film or emulsion by ionising radiation. The film, about the size of a dental X-ray film, is carried in a composite plastic holder and worn for a fixed period, usually 1–4 weeks, after which it is processed under standardised conditions and the density of blackening measured using a densitometer.

Although for fast neutron measurements it is possible with a microscope to detect and count the tracks made by individual particles, for other radiations the general density of the blackening is used as a measure of exposure. Unfortunately, because the film contains material of high atomic number, e.g. silver, the density produced by a given quantity of radiation depends on the energy of the radiation, so that a simple density—exposure relationship does not hold. To enable the energy and type of radiation to be estimated, the film is therefore partially covered with filters of different materials and thicknesses (see Fig. 14.9(a)). For example, the film under the plastic filters, when compared with that under the open window, indicates the presence of β-radiation of different energies. On the other hand, the metal filters absorb β-particles, but enable some differentiation to be made between high- and low-energy photons. For instance, the aluminium filter will stop low-voltage X-rays, but not high-voltage X-rays, whereas all X-rays suffer some attentuation in tin. When an assessment of the quality of the radiation has been made, the density measured under the open window can be corrected to give a reliable estimate of exposure.

Emulsions of different sensitivities coat the two sides of the base of the film, (Fig. 14.9(b)). If the film receives only a very small exposure, the sensitive emulsion is sufficiently blackened to permit accurate density measurements. If this emulsion is blackened too heavily for measurement, it is stripped off the base, leaving the less

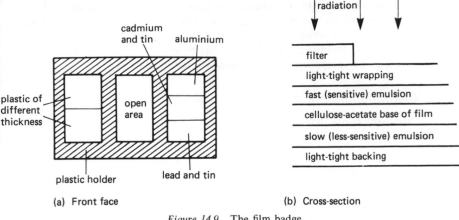

Figure 14.9   The film badge

blackened slower emulsion to provide the density measurements. In this way, the range of the film is considerably extended, so that exposures differing by factors of 10^4 can be successfully measured.

Exposures exceeding about 5×10^{-6} C kg^{-1} can be measured, any weaker β- or γ-radiation being absorbed in the film wrapper. Although the accuracy of measurement is only 10–20 per cent, this is nevertheless sufficient for protection purposes.

Characteristics

Certain features of a developed film are immediately obvious:

(a) large differences in film density suggest exposure to low-energy radiation, whilst small differences indicate high-energy radiation;

(b) X-ray exposure tends to produce sharper images than γ-ray exposure;

(c) particular patterns, (due to splashing of radioactive material, the dial of a luminous watch, and so on) can be identified.

The film badge provides a cheap, permanent record of exposure; there is no associated circuitry to buy and maintain; and the films can easily be distributed by post to a central processing laboratory. Furthermore, mixed radiation can be determined from one film.

On the other hand, there is some delay before the exposure is known, high ambient temperature and humidity can cause deterioration of the emulsion, and only radiation near the personnel or site monitor is recorded.

Thermoluminescent dosimeters

Basic principles

Certain materials, such as lithium fluoride (LiF) and calcium fluoride (CaF$_2$), when exposed to radiation, store a small fraction of the absorbed energy in metastable energy states. If the material is subsequently heated, it releases its stored energy as visible light, a phenomenon known as thermoluminescence. As in the scintillation counter, the emitted light can be detected using a photomultiplier tube, the output from which can be amplified to give a measure of exposure or exposure rate. The dosimeter is calibrated either by exposing it to known amounts of radiation or by comparing it with a standard instrument.

The thermoluminescent phosphor

There are many thermoluminescent phosphors, but only a few have the following characteristics suitable for a dosimeter:

(a) The temperature at which light energy is released should be well defined: high enough above room temperature to ensure stability, but not too high to cause heating problems. A temperature of about 200°C is found optimum.
(b) The light output should be high (good sensitivity) and approximately proportional to exposure to facilitate calibration.
(c) The response should be approximately independent of both radiation energy and exposure rate.

LiF satisfies most of these demands, and is able to detect X-, γ-, and β-radiation as well as thermal neutrons. It records in the range 10^{-7}–$10 \ \mathrm{C \ kg^{-1}}$ and is almost air- and tissue-equivalent, making direct estimates of absorbed dose possible. For this reason it is the commonest phosphor used in personnel dosimetry.

Applications

(a) Personnel and site monitoring
Two types of dosimeter are common, each being about the same size as a film badge:

(i) LiF: The LiF, with its transparent binder, is bonded to a plastic of high melting point (Fig. 14.10) with the lower area provided with holes for

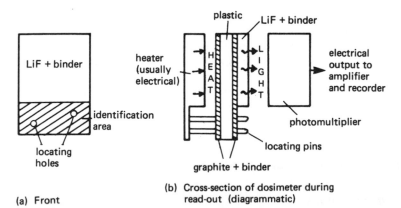

(a) Front

(b) Cross-section of dosimeter during read-out (diagrammatic)

Figure 14.10 . The LiF personnel dosimeter

positioning the dosimeter in its heating block for the read-out. This area also contains identification (person and/or location) information.

(ii) LiF and CaF_2 with manganese: The phospor mixture is incorporated in Teflon and formed into discs or tapes about 25 mm wide and 0.4 mm thick. A typical monitoring badge includes two such discs in a simple plastic retainer, one of the discs being covered by a cadmium filter. If the uncovered disc records a high reading, the other disc is uncovered and analysed to try to identify the type of irradiation.

(b) Patient exposure monitoring

Small dosimeters can be used during therapy to measure individual treatment exposures, accumulated exposures (by using the same sample during the course of treatment), and exposures for particular organs.

(c) Exposure monitoring in the body

Tiny dosimeters may be used in cavities such as the mouth, oesophagus, stomach, bladder, rectum, cervix, and uterus, or even implanted into tumours to record exposure directly.

Advantages of the thermoluminescent dosimeter include small size, high sensitivity, wide usable range, little rate or energy dependence, and good reproducibility. Production costs are reasonably low since only small quantities of LiF are required, and after use the chemical may be annealed by further heating and then re-used. However, analysis costs are high as the associated heater—photomultiplier unit is expensive, and this restricts the dosimeter's use on a small scale.

Choice of detector

The choice of detector depends on many factors including type and energy of the radiation, location of the measurement, and the cost, efficiency, accuracy and range of the detector—recording system.

In general, the Geiger counter is popular for α- and β-measurements, although liquid scintillation counting can be employed for low-energy β-radiation. Scintillation counters are almost universally used for γ-measurements, except in relatively inaccessible locations, when the thermoluminescent dosimeter is then useful. For the monitoring of low-voltage X-rays common in radiography units, monitor and thimble chambers are popular, whilst scintillation detectors are more common in the higher-energy radiotherapy units. For routine personnel monitoring, the film badge is still the preferred dosimeter.

Exercise 14

1 Describe fully an instrument used as a primary standard in the measurement of radiation exposure.

An ionisation chamber is being used to measure the activity of a source emitting 5 MeV α-particles. It may be assumed that 10 per cent of the emitted particles reach the sensitive volume of the chamber and that these are then totally absorbed in the air in the chamber. If the average ionisation energy for air at STP is 32.5 eV, and the activity of the source is 2×10^3 Bq, find the current in the chamber.

$(e = 1.6 \times 10^{-19}$ C)

2 State, giving reasons, which type of radiation detector would be most suitable for the following measurements:
   (a) accurate measurement of an X-ray machine output;
   (b) general monitoring of personnel in a radiotherapy unit;
   (c) monitoring of the radiation from ^{32}P (a pure β-emitter) introduced into the body;
   (d) γ-ray counting from a ^{60}Co source.

   The limit of sensitivity of a particular ionisation chamber is 10^{-14} A. It is used to measure the activity of a source emitting β-particles of maximum energy 1 MeV. Assuming 20 per cent of the emitted particles reach the sensitive volume of the chamber, with an average energy one third that of the maximum energy, and that they are totally absorbed there, find the minimum activity of the source, in Bq, which can be recorded.

   $(e = 1.6 \times 10^{-19}$ C; average ionisation energy for air $= 30$ eV)

3 Describe how the radiation output from an X-ray machine can be accurately measured, indicating clearly how the results are obtained from the readings taken.
   An ionisation chamber, used to monitor the exposure received by radiation workers, has a volume of 3×10^{-4} m^3, and contains air at STP. Calculate the chamber current corresponding to the maximum permissible exposure rate of 8×10^{-5} C kg^{-1} per week, assuming a five-day working week and an eight-hour working day. (Density of air at STP $= 1.29$ kg m^{-3})

4 Describe the principles of operation of the Geiger counter. In what ways does its output differ from that of a standard ionisation chamber? Give two clear examples of medical situations where the Geiger counter would be the most suitable radiation detector.
   A capacitance ionisation chamber containing air at STP is being used as an integrating dosimeter. It has a capacity of 480 pF and discharges through 14 V during an exposure of 2.6×10^{-5} C kg^{-1}. Find the volume of the chamber.
   (Density of air at STP $= 1.29$ kg m^{-3})

5 Explain, with reference to the scintillation counter:
   (a) luminescence,
   (b) scintillator,
   (c) photocathode,
   (d) dark current.
   Describe three common applications of the scintillation counter indicating briefly why it is a good choice in these situations. A NaI(Tl) crystal is found to produce 30 000 scintillations for every 1 MeV of energy absorbed in the phosphor. It is used in a scintillation counter, the conversion efficiency (photons to primary photoelectrons) of which is 15 per cent. The associated photomultiplier

tube provides an electron multiplication factor of 10^6. The counter is exposed to a ^{137}Cs source which, per 100 disintegrations, emits 85 γ-rays of energy 0.66 MeV and 10 X-rays of energy 32 keV, as well as numerous β-particles and electrons which are stopped in the counter's walls. Assuming all the γ- and X-ray energies are absorbed in the phosphor, calculate the activity of the source if a photomultiplier anode current of 10^{-8} A is obtained. ($e = 1.6 \times 10^{-19}$ C)

6 State three devices used in estimating radiation doses received by personnel using ionising radiation, and describe in detail one of them. Indicate how the type as well as the quantity of irradiation may be estimated using the selected device.

An unsealed ionisation chamber, calibrated at STP, records a reading of 5×10^{-5} C kg^{-1} per hour, when exposed to γ-radiation of energy 0.1 MeV. If atmospheric pressure is 740 mmHg and the temperature is 22°C at the time of measurement, estimate:

(a) the true exposure rate in C kg^{-1} s^{-1};

(b) the absorbed dose rate in bone, assuming the conversion factor for bone at this energy is 57 Gy per C kg^{-1}.

7 (a) Explain what is meant by *dose equivalent* and why it is important in radiation dosimetry.

Calculate the energy delivered to a person of mass 70 kg by a dose equivalent to the whole body of 30 mSv (3 rem), half being acquired from radiation of quality factor one, the remainder from radiation of quality factor three.

(b) A film badge used for personal radiation monitoring contains various filters through which radiation must pass before reaching the film. Explain how this helps in making an estimate of the dose equivalent received by the wearer of the badge.

(c) Describe the principle of operation of the thermoluminescent dosimeter.

[JMB]

15 Radioactive tracer studies

Radioactive tracers

Various radioisotopes or labelled compounds (organic compounds into which radioactive atoms have been artificially incorporated) may be introduced into the body and their fate observed by monitoring their radioactivity. Such radioactive tracers can thus yield information about parts of the body such as the blood, urine or a particular organ. Some of the more common applications are listed in Table 15.1.

The amount of tracer introduced must be small so that the system under investigation is not changed, and this demands samples of high specific activity. Furthermore, the tracer must behave like the tracee (the substance being investigated) so that it accurately reflects body behaviour. For instance, if the tracer

Table 15.1 Tracer applications

	Study	Tracers	Comments
1	Body composition	^{3}H, ^{82}Br, ^{24}Na, ^{42}K	Volumes of various body fluids and quantities of, e.g. sodium, chlorine and potassium, in the body estimated.
2	Bone	^{45}Ca, ^{47}Ca, ^{85}Sr, ^{99m}Tc	Calcium absorption, bone mineral metabolism and localisation of bone disease investigated.
3	Blood	^{131}I, ^{132}I, ^{125}I, ^{51}Cr, ^{32}P	Various volumes, e.g. plasma, red blood cell, total blood, limb blood, estimated. Internal bleeding sites located using radioactive endoradiosonde.
4	Thyroid	^{131}I, ^{132}I, ^{125}I, ^{123}I, ^{99m}Tc	Assessment of thyroid function. ^{132}I (smaller doses) useful for pregnant women and children.
5	Liver	^{198}Au, ^{131}I, ^{32}P, ^{99m}Tc,	Liver disease and disorders of hepatic circulation diagnosed.
6	Heart and lungs	^{131}I, ^{133}Xe, ^{99m}Tc	Cardiac output, blood volume and circulation times evaluated. Labelled gases used in respiration studies.
7	Tumours	^{32}P, ^{99m}Tc, ^{131}I	Detection, localisation and differential diagnosis of tumours.
8	Therapy	^{131}I, ^{32}P	Radioiodine used in thyroid treatment and radioactive phosphorus in certain haematological conditions.

is a radioisotope, the small difference in mass between it (e.g. ^{24}Na) and the tracee (e.g. ^{23}Na) is generally negligible, whereas significant effects can arise when using labelled compounds.

The tracer should be detected easily and accurately. γ-emitters are preferred, since external scintillation counters can then be used; β-emitters sometimes require the use of internal miniaturised Geiger counters, and lead to higher absorbed doses.

Both the half-life of the tracer and the biological half-life of the observed process need to be considered when evaluating the most suitable tracer and its activity, and when planning storage, handling and waste disposal.

The single most important radionuclide used at present is technetium-99m (^{99m}Tc), the exploitation of which during the last decade has been largely responsible for the rapid growth of nuclear medicine imaging techniques. Its advantages are many:

(a) It decays with a short half-life (≈ 6 hours).
(b) It emits only γ-rays.
(c) The γ-rays emitted (140 keV) are easily detected by the gamma camera.
(d) It is easily attached to many different chemical compounds for various tracer investigations.
(e) It is easily produced in situ using a 'cow' (see page 167).

Other important radionuclides are given in the following applications.

Dilution analysis

During tracer investigations of body fluids, a small quantity of tracer is introduced to, and subsequently mixes with, the volume of fluid of interest. Since the tracer is simply becoming diluted in a theoretically fixed tracee, the technique is called dilution analysis.

If the activity of the tracer administered is x Bq, and it completely mixes with the volume, Vm^3, of the tracee, the activity concentration (activity per unit volume) produced in the mixture is:

$$c = \frac{x}{V} \, \text{Bq m}^{-3}.$$

If the tracer is added as a small volume v of solution of high activity concentration c', then:

$$x = c'v$$

and
$$V = \frac{c'v}{c} \qquad\qquad [15.1]$$

Thus, if c, c' and v can be measured, V can be estimated.

Two possible sources of error in dilution analysis are:

(a) the mixing of tracer and tracee may be neither immediate nor complete;
(b) there may be some loss of tracee from the system. A graphical analysis (tracer concentration against time) may permit suitable corrections to be made.

Examples of dilution analysis

Measurement of total body water

The tracer commonly used to investigate total body water is tritiated water which is both cheap and easily detected. The relevant radioisotope is ^3H having a half-life of 12.2 years (thus decaying insignificantly during the study) and emitting β-particles of maximum energy 18 keV. The activities of samples containing the tracer are measured using liquid scintillation counters.

A known volume of tritiated water of activity about 10^4 Bq for an average 70 kg man, is given orally or injected intravenously. A similar volume of the same tritiated water is diluted with distilled water to a final volume of a few litres (10^{-3} m^3) for subsequent use as a standard. The subject is not allowed to eat or drink during the period required for tracer equilibration (complete mixing) which ranges from about 3 hours, in the absence of abnormal fluid retention, to 8 hours. A small blood sample is then taken and centrifuged to separate the plasma.[1] The activities of equal volumes of the plasma and diluted standard are then measured by liquid scintillation counting, yielding count rates of p and s respectively. (These readings should be of comparable values, permitting them to be taken on the same range of the meter.)
Then:

$$\frac{s}{p} = \frac{\text{activity of diluted standard sample}}{\text{activity of plasma sample of equal volume}}$$

$$= \frac{c'/d}{c}$$

where c' = original activity concentration of the undiluted tracer (^3H)

d = dilution factor of the standard (ratio of diluted to undiluted volume)

c = final activity concentration of the plasma

$$\therefore \quad \frac{c'}{c} = \frac{ds}{p}$$

Substituting into equation [15.1] gives the volume of total body water as:

$$V = \frac{dsv}{p} \qquad [15.2]$$

where v is the volume of tracer administered. For the average 70 kg male, V is approximately 0.04 m^3 and agreement to within 3 per cent is obtained in repeated measurements using this technique.

Measurement of plasma volume

Plasma proteins such as albumin, fibrinogen and γ-globulins have been success-fully labelled with radioiodine (^{125}I, ^{131}I or ^{132}I) and, provided no more than one

[1](Plasma is the almost colourless liquid of the blood stream, in which float the red and white corpuscles and platelets. Here it is considered as part of the total body water.)

atom of the radioisotope is introduced per protein molecule, the biological properties of the proteins seem unaffected. Since the iodine is bound to a comparatively large protein molecule, it cannot readily escape from the circulatory system or be absorbed by the thyroid gland and so it is a good indicator for the blood stream.

Human serum albumin labelled with ^{131}I is the commonest tracer used for plasma volume determinations. ^{131}I has a half-life of about 8 days and its most intense γ-radiation is at an energy of 0.36 MeV. The tracer used is in fact an aqueous solution of labelled albumin, thoroughly mixed with saline solution and sodium iodide solution.

A suitable volume (of activity about 10^5 Bq) of the tracer solution is injected intravenously, whilst some of the same solution is accurately diluted with distilled water to serve as a standard. A blood sample is taken fifteen minutes later from a different vein, and preferably a different limb, from that which was injected, to exclude contamination from the injection site. The blood sample is centrifuged to separate the plasma and the activities of equal volumes of plasma and diluted standard are measured in a well-type scintillation counter. The plasma volume is then estimated using equation [15.2]:

$$\text{plasma volume} = \frac{dsv}{p}$$

where d = dilution factor of the standard

   s = count rate of the diluted standard

   v = volume of labelled albumin solution injected

   p = count rate of the plasma sample.

A correction to this estimate is necessary to allow for the loss of activity during the tracer equilibration period, which may approach 20 per cent per hour for patients with burns but which is normally less than 10 per cent per hour. In addition, corrections have to be made to the count rates to allow for background radiation, a corrective procedure necessary in all such examinations. A normal corrected value for the plasma volume of an average 70 kg man is 3×10^{-3} m^3 (3 litres).

Measurement of red blood cell volume

^{51}Cr is commonly used for labelling red blood cells. It has a half-life of 27.8 days and decays by electron capture emitting γ-rays of energy 0.322 MeV, as well as characteristic X-rays.

Sodium chromate solution, labelled with about 10^6 Bq of ^{51}Cr, is mixed with a red cell suspension, derived from a small sample of the patient's own blood, and left at 37°C for about thirty minutes. During this time, the ^{51}Cr attaches itself to the red blood cells. The labelled red cells are then carefully washed with plasma saline to remove any unattached radioactive chromate and then further diluted with plasma saline. A known volume of this tracer is injected into the patient, and a similar volume is diluted with unlabelled sodium chromate solution and used as a standard.

A blood sample is removed from a remote vein after about fifteen minutes and a well-type scintillation counter is used to measure the activities in identical volumes of the diluted standard (count rate, s) and the blood sample, (count rate, b).

Total blood volume V_T is then given, using equation [15.2] by:

$$V_T = \frac{dsv}{b}$$

where d is the dilution factor of the standard, and v is the volume of the labelled red cells injected.

V_T consists of two components, namely the volume of red blood cells V_R and the plasma volume V_P. Thus:

$$V_T = V_R + V_P.$$

In addition, a term known as the haematocrit, H, is defined as the ratio:

$$H = \frac{V_R}{V_T} \times 100\% \qquad\qquad (\approx 45\%, \text{ normally})$$

H may be estimated by centrifuging the blood sample, so that the red cells gather together at the bottom of the tube. Then:

$$V_R = V_T \times \frac{H}{100}$$

$$\therefore \qquad V_R = \frac{dsv}{\cdot b} \times \frac{H}{100}$$

V_R can thus be estimated as can V_P also, if required. A typical result for the red cell volume of an average 70 kg man would be $2 \times 10^{-3} \, m^3$. (2 litres)

Measurement of cardiac output

Tracer analysis of circulating fluids

Radioactive tracers may be used not only to investigate blood flow, yielding information about circulation times and blood volumes in large or localised areas, but also to study gas flow during respiration, using the various radioactive gases available.

In blood flow studies, a known amount of tracer is injected into the circulation, so that it mixes with the blood flow to be measured, and its subsequent concentration either at, or downstream of, the mixing site is monitored, often using an external radiation detector. A plot of the resulting activity against time produces a tracer concentration curve.

If the activity of the tracer administered is x Bq and its mean blood concentration during the first passage is y Bq m^{-3}, then the volume V of the tracee is:

$$V = \frac{x}{y} \, m^3$$

Further, if the duration of the first passage is t, then the flow rate is:

$$\frac{V}{t}=\frac{x}{yt}\ \text{m}^3\,\text{s}^{-1}.$$

Cardiac output

This simple analysis may be used to estimate cardiac output, Q (the volume of blood pumped per second) if the following assumptions are made:

(a) the tracer mixes with the whole of the cardiac output;
(b) there is no loss of tracer between injection and sampling point;
(c) the entire first passage of the tracer is identified with no contribution from recirculating tracer.

Then, Q is evaluated using:

$$Q=\frac{V}{t}=\frac{x}{yt}\ \text{m}^3\,\text{s}^{-1}.$$

Radiocardiography (RCG)

Radiocardiography is the technique of monitoring the radioactive tracer as it flows through the heart chambers, and the resulting tracer concentration curve is referred to as a radiocardiogram. Figure 15.1 illustrates a typical radiocardiogram consisting of two close peaks corresponding to the passage of the tracer through the right and left chambers respectively, while the dip in between corresponds to the tracer's passage through the lungs which are generally not in the detector's field of view. In order to separate the contribution made by the recirculating tracer from the tail of the first-passage curve, the latter is extrapolated along an exponential downslope as indicated by the dotted line in the figure.

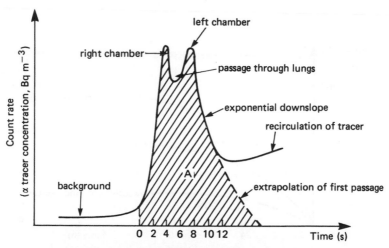

Figure 15.1   A typical radiocardiogram (RCG)

^{131}I-labelled human serum albumin is a popular tracer for cardiac studies. A solution of about 10^6 Bq of the labelled plasma protein in saline is injected, preferably into the superior vena cava whence it is transported into the right auricle (see Fig. 5.4). The detector, a collimated NaI (Tl) scintillation counter, should be positioned over the centre of the heart image outlined by fluoroscopy. The cardiac output is then estimated using:

$$Q = \frac{x}{yt} \text{ m}^3 \text{ s}^{-1}$$

Since y is the mean blood concentration of the tracer and t the duration of the first passage the product yt is equivalent to the area A (Bq m^{-3} s) under the curve (shown shaded).

$$\therefore \qquad\qquad Q = \frac{x}{A} \text{ m}^3 \text{ s}^{-1}$$

A should be corrected for background counts.

Radiocardiographic estimates of cardiac output agree well with those of other established methods, an average result being $Q \approx 10^{-4}$ m^3 s^{-1}.

Assessment of thyroid function

Iodine is readily absorbed from the gastrointestinal tract into the blood stream, whence much of it is collected by the thyroid for use in the manufacture of thyroid hormones. The functioning of the thyroid may be investigated using thyroid uptake tests which involve the measurement of the accumulation of radioiodine (the tracer) by the thyroid gland.

The most common radioisotope of iodine used is ^{131}I since it is readily available using reactor methods (see page 171), it has a convenient half-life of 8.06 days and its γ-radiation (0.364 MeV) is easily detected using scintillation counters. The patient is given a tracer dose of about 10^6 Bq, normally orally in the form of a dilute

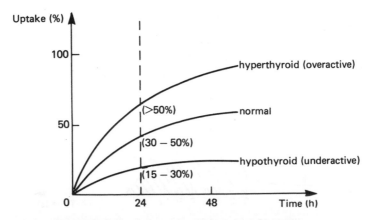

Figure 15.2   Uptake of radioiodine by the thyroid gland

sodium iodide solution, but sometimes intravenously. The count rate from the thyroid, measured using a suitably collimated NaI (Tl) scintillation counter, is then compared with that from a standard containing the same amount of ^{131}I as in the administered dose and having a volume comparable to that of an average thyroid gland, ($\approx 3 \times 10^{-5}$ m^3). The percentage uptake, given by:

$$\frac{\text{count rate from thyroid}}{\text{count rate from standard}} \times 100\%$$

is evaluated after specified time periods, from 10 minutes to 48 hours. Typical values for the 24-hour test (perhaps the most common) are indicated on Fig. 15.2.

Localisation studies

General principles

Certain isotopes have a tendency to settle in particular parts of the body; for example:

(a) iodine accumulates in the thyroid;
(b) calcium and strontium isotopes are concentrated in bone;
(c) potassium and rubidium have an affinity for muscle.

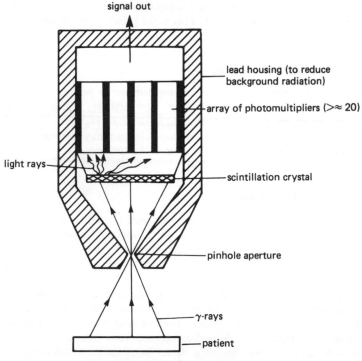

Figure 15.3 Gamma camera

If a radioisotope with such a particular affinity is administered to the body, its subsequent localisation in the selected organ may be investigated using radioactive scanning techniques. The resulting 'visualisation' of the organ can then lead to diagnoses of disorders.

Radiation scanners

A common scanning system is a moving detector system. The detector is mounted on a carriage, which moves automatically across the patient, scanning the area of interest and simultaneously feeding the results to a synchronised display system. The NaI (Tl) scintillation counter is a popular detector for such scanning systems.

Alternatively, a fixed detector system, such as the gamma camera or autofluoro-scope, may be employed. In the gamma camera (Fig. 15.3), radiation from the patient is collimated by a small hole in the lead housing, and produces scintil-lations in a large but thin NaI (Tl) crystal, the image crystal. The coordinates of each such scintillation are defined in terms of correlated signals from an array of photomultiplier tubes mounted on the back face of the crystal. Each scintillation is then caused to produce a light flash on a cathode ray tube screen at a position corresponding to its position in the image crystal. An image duplicating the activity distribution in the patient is thus built up and recorded, usually on Polaroid film.

If such a scintillation camera is linked to a closed-circuit television system, it is possible to observe the progress of a radioactive tracer in its passage through a section (e.g. heart, brain) of the body. This is the principle of the autofluoroscope.

Thyroid localisation studies

The thyroid may be visualised using a moving detector or gamma camera scanning system. Any 'hot' (high activity) or 'cold' (low activity) regions may then be identified. Goitres (enlargement of the thyroid gland) and other abnormalities in shape, size, and position of the gland can be diagnosed, as also may tumours which do not concentrate iodine and hence appear as cold or inactive areas.

Bone localisation studies

Various bone-seeking radioisotopes have been used in studies of bone metabolism and the localisation of bone disease, including notably isotopes of calcium and strontium. Whereas ^{45}Ca emits weak β-particles and ^{47}Ca emits high-energy γ-rays (1.3 MeV), ^{85}Sr emits moderate-energy (0.51 MeV) γ-rays and is thus a popular tracer.

After an intravenous injection of ^{85}Sr, the normal skeleton displays a rapid increase in count rate to a peak value after 24 hours, followed by a gradual and slow decrease. Any bone abnormalities tend to lead to an accumulation of the tracer, resulting in high and prolonged count rates from these areas.

Bone localisation studies can often detect bone tumours, fractures, infections and osteoarthritis before these become evident using radiographic techniques.

Exercise 15

1 Discuss the factors which affect the choice of radioisotope for use in a particular tracer study.

Give three clinical examples of the use of radioactive tracers, and in each case include:
(a) the reasons for the choice of tracer,
(b) the method of detection of the tracer.

The radiation-absorbed dose delivered to a patient during a total body water measurement is found to be 2×10^{-11} Gy per Bq of initial activity administered. Using the following data, calculate the absorbed dose received by the patient.

$$\text{Volume of tracer administered} = 5 \times 10^{-6} \, \text{m}^3$$
$$\text{Dilution factor of standard} = 400$$
$$\text{Activity of } 20 \times 10^{-6} \, \text{m}^3 \text{ of diluted standard} = 100 \, \text{Bq}.$$

2 Describe a radioactive tracer study in which the tracer:

(a) mixes with the substance under investigation;
(b) is accumulated in the object of interest.

Using the following data, estimate the absorbed doses received during the two examinations:
(c) Absorbed dose per Bq of tracer administered $= 1.7 \times 10^{-11}$ Gy per Bq.

$$\text{Volume of tracer administered} = 8 \times 10^{-6} \, \text{m}^3.$$
$$\text{Volume of diluted standard} = 30 \times 10^{-6} \, \text{m}^3.$$
$$\text{Activity of diluted standard} = 200 \, \text{Bq}.$$
$$\text{Dilution factor of standard} = 800.$$

(d) Absorbed dose per Bq of tracer administered $= 8.6 \times 10^{-10}$ Gy per Bq.

$$\text{Volume of tracer administered} = 5 \times 10^{-6} \, \text{m}^3$$
$$\text{Activity concentration of tracer} = 10^{11} \, \text{Bq m}^{-3}.$$

3 Explain briefly how a radiocardiogram may be obtained. Describe its essential features and show how it may be used to provide an estimate of cardiac output.

During a cardiac output study, 2.5×10^{-7} m^3 of tracer solution is injected and a radiocardiogram obtained. The area under the first-passage curve is estimated to be 10^{10} Bq m^{-3} s. A similar volume of the same tracer solution is diluted and used as a standard. It is found that a sample of 10^{-6} m^3 of the standard has an activity of 8×10^4 Bq. If the dilution factor of the standard is 100, estimate the cardiac output.

4 Give an account of the method of dilution analysis for measuring various body fluid volumes, giving particular attention to any assumptions or possible sources of error in such studies.

The following results were obtained during a radioactive tracer investigation of red blood cell volume:

$$\text{Volume of labelled red cells injected} = 8.83 \times 10^{-6} \, \text{m}^3$$
$$\text{Dilution factor of standard} = 100$$

Count rate of diluted standard (3 readings) $= 167.4 \text{ s}^{-1}$
166.2 s^{-1}
166.8 s^{-1}
Count rate of venous blood sample $= 37.2 \text{ s}^{-1}$
(3 readings) 38.1 s^{-1}
37.7 s^{-1}

If the red blood cell volume was found to be $1.78 \times 10^{-3} \text{ m}^3$, estimate the haematocrit of the venous blood sample.

5 Radioiodine is used in the tracer studies of both plasma volume and thyroid uptake. Why does the thyroid function not interfere with the plasma volume studies?

   Describe how the measurement of plasma volume is made, indicating clearly how the result is obtained from the readings taken.

   During such a tracer study of plasma volume the following results were obtained:

$$\text{Volume of tracer injected} = 8.66 \times 10^{-6} \text{ m}^3$$
$$\text{Dilution of standard} = 10^{-6} \text{ m}^3 \text{ in } 10^{-4} \text{ m}^3$$

Count rate of $2 \times 10^{-6} \text{ m}^3$ of diluted
standard (3 readings) $= \begin{cases} 338.8 \text{ s}^{-1} \\ 332.0 \text{ s}^{-1} \\ 335.4 \text{ s}^{-1} \end{cases}$

Count rate of $2 \times 10^{-6} \text{ m}^3$ of plasma
(3 readings) $= \begin{cases} 90.45 \text{ s}^{-1} \\ 90.1 \text{ s}^{-1} \\ 90.05 \text{ s}^{-1} \end{cases}$

Use the data to estimate the plasma volume.

6 (a) Give two examples of radioisotopes which have been used for diagnostic purposes in medicine. State how one of these radioisotopes may be produced.
   (b) Two samples, each of volume 10^{-6} m^3, contained tritiated water, the activity of each sample being 4.00 MBq ($108 \, \mu\text{C}$). One sample was injected into the bloodstream of a patient and after a suitable period of time $4.00 \times 10^{-6} \text{ m}^3$ of blood was withdrawn. The corrected count rate produced by this blood in a liquid scintillation counter was 207 counts per second. The other sample was diluted 10 000 times with ordinary water and $4.00 \times 10^{-6} \text{ m}^3$ of the diluted liquid produced 745 counts per second, after correction, in the same scintillation counter.

   Calculate the volume of water in the patient's body. The half-life of tritium may be assumed to be very long compared with the duration of the experiment.
   (c) Explain, for the measurements in (b),
       (i) the nature of the correction applied to the count rates,
       (ii) why the volume of fluid injected into the bloodstream needed to be small,
       (iii) why time was allowed to elapse before withdrawing the blood for measurement,
       (iv) why, apart from convenience of calculation, the dilution factor was chosen to be 10 000.

[JMB]

Answers

Exercise 1

1 12 (to nearest person)
3 0.083 W; 1.67 W; 1.96%; 7.84%
4 (a) 57.2 W (b) 381 W (c) 621 W (d)564 W (e)5
6 42

Exercise 2

1 (a) 720 N; (b) 590 N
2 193 N; 230 N; 268 N; 225 N
3 $\frac{1}{3}$ way along vertebral column from top; (a) 1037 N (b) 2332 N
4 (a) 43.75 N (b) 29 m
5 (a) $3.2\,\mathrm{m\,s^{-1}}$ (b) 1.612 m (c) 358 J (d) 1434 W
7 (e) 1064 N (f) 364 N; G = 1155 N; F = 577 N
8 (c) $6.32\,\mathrm{m\,s^{-1}}$ (d) 0.126 s (e) 3500 N (f) $5\,\mathrm{MN\,m^{-2}}$; No; New
   compressive stress $= 2 \times 10^8\,\mathrm{Nm^{-2}}$, therefore fracture

Exercise 3

1 (c) 7.3 mm
2 (a) 0.073 mm (b) converging lens of $f = 286$ mm; range is 250 to 353 mm
5 1.83 D
6 (a) far-sighted; converging lens of $f = 375$ mm
   (b) defective green-sensitive receptors (deuteranopia)
   (c) near-sighted; diverging lens of $f = -2$ m
   (d) astigmatism; cylindrical lens
   (e) presbyopia; converging lens of $f = 500$ mm
7 (a) diverging lens of $f = -200$ mm; new range is 200 mm to infinity
   (b) far point at 2 m; near point at 500 mm
8 (a) range is 600 mm to infinity
   (b) range is 250 mm to 429 mm
11 (b) 4 D; concave lens of power 1 D; 176.5 mm

Exercise 4

1 3.5 dB
2 0.999 (to 3 decimal places)
3 (a) $10^{-3}\,\mathrm{W\,m^{-2}}$ (b) 90 dB (c) 26 dB
4 (c) 0.969 dB (d) 99%

5 (a) 6.02 dB   (b) −9.54 dB   (c) 23.0 dB
6 (a) 2.48 dB   (b) 2.99 dB
7 (i) Intensity level increases by 0.969 dB, 1.072 dB, 0.969 dB (i.e. approximately by 1 dB each time)   (ii) 1.12
8 30 m
9 (a) 1 W m^{-2} (b) 0.597
10 (b) 5.01 mW m^{-2}, 10.02 mW m^{-2}, 3 dB
11 (b) $1.5 \times 10^{-4} \text{ m}$

Exercise 6

2 133 W
3 $0.00146 \text{ K sec}^{-1}$; 5.25 K; No
4 31.6°C
5 (e) 30.1 W   (f) 113 W
6 (b) (i) 122.4 J min^{-1} (ii) 607.2 J min^{-1}

Exercise 7

5 0.197%
6 (a) Increased by factor of 10   (b) 27.2 kPa; 6.37 kPa

Exercise 9

3 (a) Angle of incidence = 76.8°, critical angle = 69.0°, therefore total internal reflection occurs
  (b) 7.86°
4 (d) 0.515   (e) 31°
5 28.8%
7 From 53 MW m^{-2} to 229 GW m^{-2}

Exercise 10

1 (d) 99.89%   (e) 0.446%   (f) 99.88%
2 $\alpha_{\text{quartz/tissue}} = 0.646$; $\alpha_{\text{quartz-perspex}} = 0.420$; $\alpha_{\text{perspex-tissue}} = 0.106$
  ∴ quartz–tissue transmission = 35.4%
  quartz–perspex–tissue transmission = 51.8% (Better)
4 (a) 33%   (b) 31.5%
5 (d) 96.5%   (e) 95.4%
7 0.15 ms^{-1}
8 $0.5 \times 10^{-6} \text{ m}^3 \text{ s}^{-1}$

Exercise 11

2 $1.12 \times 10^{-14} \text{ J}$; $1.77 \times 10^{-11} \text{ m}$
5 0.353 mm
6 (a) $3.75 \times 10^{17} \text{ s}^{-1}$ (b) 5.94 kW
7 (a) 30 MW m^{-2} (b) 7.5 MW m^{-2}
11 (a) (i) B   (iv) 1 mm

Exercise 12

1 $9.63 \times 10^{-5} \, \text{s}^{-1}$
2 $4.66 \times 10^{18} \, \text{Bq}$
3 19 100 years
4 8100 years
5 (c) 218; 84   (d) $1.09 \times 10^{-19} \, \text{kg m s}^{-1}$ (e) $307 \, \text{km s}^{-1}$
6 (a) $1.54 \times 10^{12} \, \text{Bq}$ (b) 60 MJ
7 $6000 \, \text{cm}^3$
8 (a) 2   (b) 1   (c) $\frac{1}{8}$

Exercise 13

1 6.60×10^{16}
2 33.2 mm
3 $4.61 \times 10^{15} \, \text{Bq}$
4 (b) no shield   (c) use longer tongs
5 (b) 34 Gy
7 (b) 0.63

Exercise 14

1 $4.92 \times 10^{-12} \, \text{A}$
2 28 Bq
3 $2.15 \times 10^{-13} \, \text{A}$
4 $2 \times 10^{-4} \, \text{m}^3$
5 24.6 Bq
6 (a) $1.54 \times 10^{-8} \, \text{C kg}^{-1} \text{s}^{-1}$ (b) $8.79 \times 10^{-7} \, \text{Gy s}^{-1}$
7 (a) 1.4 J

Exercise 15

1 $2 \times 10^{-7} \, \text{Gy}$
2 (c) $7.25 \times 10^{-7} \, \text{Gy}$ (d) $4.3 \times 10^{-4} \, \text{Gy}$
3 $2 \times 10^{-4} \, \text{m}^3 \, \text{s}^{-1}$
4 45.5%
5 $3.22 \times 10^{-3} \, \text{m}^3$
6 (b) $0.036 \, \text{m}^3$

Bibliography

E. H. Belcher and H. Vetter: Radioisotopes in Medical Diagnosis. Butterworths, 1971.

G. B. Benedek and F. M. Villass: Physics: With Illustrative Examples from Medicine and Biology, Volume 1. Addison-Wesley, 1974.

J. Blitz: Fundamentals of Ultrasonics. Butterworths, 1967.

E. J. Casey: Biophysics: Concepts and Mechanisms. Reinhold, 1962.

V. H. Frankel et al.: Orthopaedic Biomechanics. Lea and Febager, 1970.

A. C. Guyton: Textbook of Medical Physiology. W. B. Saunders, 1971.

Michael Maisey: Nuclear Medicine, A Clinical Introduction. Update, 1980.

W. J. Meredith and J. B. Massey: Fundamental Physics of Radiology. John Wright & Sons, 1972.

W. L. Nyborg and M. C. Ziskin: Biological Effects of Ultrasound. Churchill Livingstone, 1985.

R. Oliver: Radiation Physics in Radiology. Blackwell, 1966.

P. R. Salmon: Fibre-optic Endoscopy. Pitmans, 1974.

Scientific American: Medical Thermography. 216 p. 94, 1967.

K. J. W. Taylor et al.: Manual of Ultrasonography. Churchill Livingstone, 1980.

D. C. S. White: Biological Physics. Chapman and Hall, 1974.

M. Williams and H. R. Lissmer: Biomechanics of Human Motion. W. B. Saunders, 1962.

H. S. Wolff: Biomedical Engineering. World University Library, 1970.

Index

Heinemann Educational Books Ltd
Halley Court, Jordan Hill, Oxford OX2 8EJ

OXFORD LONDON EDINBURGH MADRID ATHENS BOLOGNA
PARIS MELBOURNE SYDNEY AUCKLAND SINGAPORE TOKYO
IBADAN NAIROBI HARARE GABORONE PORTSMOUTH NH (USA)

ISBN 0 435 68682 8

First published 1984 Second edition 1989
94 95 96 13 12 11 10 9 8 7 6

Typeset by Advanced Filmsetters (Glasgow) Ltd
Printed and bound in Great Britain by
Athenaeum Press Ltd, Newcastle upon Tyne